Shale Oil and Gas: The Promise and the Peril

Vikram Rao

RTI Press

First edition published in 2012.
Second edition 2015.

The RTI Press mission is to disseminate information about RTI research, analytic tools, and technical expertise to a national and international audience. RTI Press publications are peer-reviewed by at least two independent substantive experts and one or more Press editors.

RTI International is an independent, nonprofit research organization dedicated to improving the human condition by turning knowledge into practice. RTI offers innovative research and technical services to governments and businesses worldwide in the areas of health and pharmaceuticals, education and training, surveys and statistics, advanced technology, international development, economic and social policy, energy and the environment, and laboratory testing and chemistry services.

Library of Congress Control Number: 2015946948

ISBN 978-1-934831-07-6
(refers to print version)

RTI Press publication No. BK-0012-1508
http://dx.doi.org/10.3768/rtipress.2015.bk.0012.1508
www.rti.org/rtipress

Cover design: Tayo Jolaoso
Pen-and-ink illustrations: Gordon Allen

This publication is part of the RTI Press Book series.

RTI International
3040 East Cornwallis Road, PO Box 12194, Research Triangle Park, NC 27709-2194 USA
rtipress@rti.org
www.rti.org

Dedication

To my grandmother, Srimati Hattiangadi Manorama Bai, a pioneer educator in South India. She would have been pleased.

Contents

continued

Contents *(continued)*

Figures

Preface to the Second Edition

"You can check out, but you can never leave"
—From "Hotel California" by The Eagles (written by Don Henley, Don Felder, and Glenn Frey)

This second edition was prompted by a singular event: shale oil burst onto the hydrocarbon scene with even more suddenness than did shale gas. The first edition gave it scant attention, and that needed correcting. Some may argue that now shale oil is the story. Unlike gas, oil is a world commodity. Consequently an unpredictably large new source has the potential to affect world prices. During the two-year period 2012 to 2014, new US oil production of about 2 million barrels per day (MM bpd) offset a commensurate drop in production from the rest of the world, likely helping keep prices stable. In late 2014 Libya commenced adding about 700 MM bpd to the supply, and at the same time consumption leveled off at 92 MM bpd. OPEC refused to decrease production to prop up prices, and the US continued to add to the oversupply situation. In early 2015 the price of oil dropped to $50 per barrel in a matter of months. It is safe to conclude that US shale oil is directly impacting the price of oil.

The fact that this new oil has limited domestic consumptive value because it is too good (see chapter 12 for an explanation) has made the allowance of export reasonable despite the fact that the country continues to be a net importer. Because the pipeline infrastructure was unprepared for the increase in production, a million barrels a day (and growing) have had to be transported by truck and train. There are limited outlets for the gas associated with this light oil, resulting in the flaring (wasting) of $1 billion worth of gas annually. The sudden bounty of shale oil is bringing with it some baggage.

Over the last two years, many states have moved to strengthen regulations to ensure responsible production. As of mid-2015, federal regulations do not address flaring of gas associated with oil. However, North Dakota has stairstep state regulations that are designed to minimize flaring. One area that has received much scrutiny is that of public disclosure of chemicals used for fracturing. I discuss this contentious issue in some detail in chapter 6. Chapter revisions include the most recent studies of the possibility of aquifer contamination; so far, the news is good. Since oil and gas formation have precisely the same mechanism, the early chapters underwent little change, but most chapters have been updated for recent information.

This edition has renewed support for using natural gas to displace oil as the raw material for fuels and chemicals, although a sustained low oil price could affect the degree to which this happens. I report also on some new ideas being developed to augment conventional processes (using economies of scale) with modular processes relying on economies of mass production Singly these modules can also address the monetization economically or logistically of stranded gas. In aggregate they would be alternatives to large conventional plants. One note of caution on all the economics discussions. Commodity prices have been volatile in 2014 and 2015. Where calculations are used to support any particular argument, the prices used have been spelled out. Readers can recalculate based on then-current pricing if they wish. In general, I have attempted to use prices that I consider "normal." Finally, I discuss some concepts for high-efficiency vehicles utilizing alternative fuels. In all I added six chapters.

I received a lot of feedback asking for more international emphasis. I did point out in the first edition that, based on geology, one would expect the potential for shale oil and shale gas in many other countries. But actual progress has been slow to date. Also, I expect the considerations to be the same as in the US except for the additional issue of property owners not owning mineral rights in these countries. This last issue will require the respective governments to offer property owners some inducements such as revenue sharing. I consider Argentina to be best positioned of all countries from the standpoint of prospect quality and size. Accordingly this book has a new chapter on the possibilities there. Prospective shale-extracting countries would do well to follow the US experience closely and strive to make a virtue of being late.

Glossary

Alkanes. As used in this book these are a family of compounds all with the formula C_nH_{2n+2}. The simplest alkane is with $n = 1$, which is methane. These are sometimes referred to as linear hydrocarbons.

Aquifers. Underground bodies of water. Freshwater aquifers, through drilled water wells, are the primary source of water for human consumption other than surface water.

bbl. Barrel.

bcf. Billion cubic feet.

Brackish. Water with soluble salts such as chlorides in greater concentration than fresh water (500 parts per million) but less than sea water (35,000 parts per million).

Breaker. A chemical used to eliminate the effects of the crosslinker, thus allowing the viscous fluid to be retrieved from the reservoir more easily.

Brent. Brent crude is a major trading classification of light sweet crude oil that serves as a benchmark in oil pricing. This is one of the benchmarks in oil pricing and comprises light sweet oil with an API gravity of about 38.06. See also WTI.

BTEX. Literally stands for benzene, toluene, ethylbenzene, and xylene. These are volatile organic compounds that are designated as *aromatics* because of their chemical structure. This term was originally coined because these compounds tend to have an aroma—and not a pleasant one.

CERA. Cambridge Energy Research Associates. One of the most reputable energy consulting houses, best known for their annual conference in the spring.

CNG. Compressed natural gas. Gas in this form is stored at about 3,000 psi to 3,500 psi and used for a variety of purposes.

Crack spread. The price difference between the original fluid and the final cracked fluid. After processing costs, this is reflective of the profit in the operation.

Cracking. A refining process for converting larger molecules to smaller, more useful ones. This can be done thermally and/or by using specialized catalysts.

Crosslinker. A chemical used in some fracturing operations to make the fracturing fluid more viscous. Doing so increases the efficacy of producing fractures in the rock.

Crude oil. A mixture of naturally occurring hydrocarbons which can be processed in refineries to yield useful fluids such as gasoline, diesel, and jet fuel. The bulk of crude oil has the formula C_nH_{2n+2}, where n is generally over 20.

CTL. Coal-to-liquids. This term describes the process for converting coal to liquid hydrocarbons. The original application of the F-T process was CTL.

Diesel. A crude oil distillate that can be compression-ignited, as opposed to spark ignition, required for gasoline.

DME. Dimethyl ether. This is a simple ether derived from methanol. It can be blended with diesel to at least 20 percent by volume without engine modifications. It has a higher cetane rating than diesel and produces zero particulates when combusted.

Drop-in fuel. This is a class of fuels which may be blended with, or "dropped into," conventional automotive fuels. In the pure definition, this ought to be possible in any proportion.

Dry gas. Natural gas that is substantially free of NGLs.

EIA. The Energy Information Administration is a section of the US Department of Energy (DOE) that collects, analyzes, and disseminates statistics and data on resources, supply, production, and consumption for all energy sources.

FFV. Flex-fuel vehicle. This is a vehicle with an engine that can operate on any mix of gasoline and alcohol.

Fischer-Tropsch. Abbreviated F-T or FT, this is one of the most common gas-to-liquid processes.

Flocculent. A fluffy material which, in the context of water treatment, captures undesirable species and is then discarded.

Flowback water. Water used in the fracturing operation that is circulated back to the surface after the operation is complete. It often contains some proportion of water from the rock formation. In shale oil and gas operations, the flowback water is decidedly more salty than what went in. This is distinct from *produced water*, which is formation water that tends to flow after the hydrocarbon is mostly extracted. They do overlap, in that the flowback water will have a component of formation water.

Formation water. The natural water layer in a natural gas or coal reservoir.

Fossil fuel. Fuel consisting of the remains of organisms preserved in rocks in the earth's crust with high carbon and hydrogen content.

Fracturing. Defined as an operation in which high-pressure fluid, usually water- based, is injected into the reservoir rock in order to fracture it to induce artificial permeability.

F-T. See Fischer-Tropsch.

Fugitive gas emissions. Methane released inadvertently into the atmosphere or into a body of water.

Gasoline. A crude oil distillate that powers internal combustion engines; can also be synthesized from natural gas.

GTL. Gas-to-liquids. This term stands for the process of converting natural gas to useful liquids and is usually used in connection with producing transport fuel.

Henry Hub price. This is the price in Erath, Louisiana, for natural gas, and is used for trading on the New York Mercantile Exchange. It is strongly correlated with the price of gas in all parts of the US.

Interval (well). A well interval is a portion along the length of the well. The designation is used to identify zones of different character, such as productivity.

IHS. A large consulting house of which CERA (Cambridge Energy Research Associates) is currently a part.

Kerogen. A mixture of long-chain organic molecules that is a precursor of oil and gas.

LNG. Liquefied natural gas. This is natural gas chilled to the fluid state and kept chilled at about -260°F. Its volume is 600 times less than free gas. Transport of natural gas across the ocean is done in LNG tankers.

LPG. Liquefied petroleum gas. This is a mixture of molecules larger than ethane and methane. Most LPG is some blend of propane and butane.

Marcellus. The name of the formation of sedimentary rock in the eastern United States containing important deposits of shale gas.

mcf. Thousand cubic feet.

Methane. A colorless, odorless, flammable gas with the formula CH_4 and the principal constituent of natural gas.

MM bpd. Million barrels per day.

MM Btu. Million British thermal units. This is the most common energy unit to describe the energy content in fossil fuels. A thousand cubic feet of natural gas nominally contains 1 million Btu.

MM cfd. Million cubic feet per day.

MTG. Methanol to gasoline. This is a specific GTL process which converts syngas to methanol and then to gasoline. The syngas may be from any carbonaceous source but usually comes from natural gas.

Naphtha. This is product of oil distillation during a refining operation. It is defined as the fraction of hydrocarbons boiling between 30°C and 200°C. It consists of a complex mixture of hydrocarbon molecules generally having between 5 and 12 carbon atoms in the formula C_nH_{2n+2}. It is the starting point for a number of useful fuels and chemicals.

Natural gas. Natural gas is a hydrocarbon primarily comprising methane but often mixed with larger molecules in the alkane family.

NGL. Natural gas liquids. These are molecules larger than methane found in association with methane in natural gas. In ascending order of size they are ethane, propane, butane, and larger molecules beyond.

Oil shale. A shale rock infused with immature oil and kerogen that can be thermally processed to yield useful hydrocarbons. This is not to be confused with *shale oil*, which is separately defined below.

Pad drilling. This is a relatively new type of drilling wherein anywhere from 5 to 40 wells may be drilled from a single location, or "pad." The advantages include less overall road construction and traffic. It also facilitates water treatment operations due to economies of scale.

Permeability. The property of a rock that defines ease of mobility of fluid through it.

Petroleum. The technically exact definition is a mixture of naturally occurring hydrocarbons that includes oil and gas. The petroleum industry encompasses both fluids. However, in the parlance *petroleum* is more commonly used synonymously with oil, which is liquid at room temperature.

Play (gas play). A region being mined for petroleum, or the activities associated with its development.

Produced water. This is formation water that usually flows out after much of the hydrocarbon has been produced. It is distinct from flowback water, but a portion of flowback water is formation water. Industry nomenclature is not clear on this point, and sometimes flowback water is referred to as "produced water."

Proppant. A hard ceramic material, usually sand, which is placed in the fractures created in the reservoir rock. The purpose is to "prop" the fractures open to allow gas or oil to flow. Absent the proppant, the weight of the thousands of feet of rock above would close the fractures.

Refracturing. The process of fracturing a reservoir a second time in an existing well bore. This is usually conducted a few years after producing gas or oil from the original fractures.

Reserves estimate. This is an estimate of the hydrocarbon resource which is economically recoverable using current technology. In most cases this estimate will increase as the resource is developed, especially in new plays such as shale gas.

Resource estimate. This is an estimate of the hydrocarbon likely to be in place, whether economically recoverable or not.

RPSEA. Research Partnership to Secure Energy for America. This organization was formed with congressional line-item funding and includes most players in the oil and gas business. Initial targets were ultra-deepwater and unconventional resources.

Saline aquifers. Aquifers with water too salty for human consumption, generally defined as having greater than 500 parts per million chlorides. Saline aquifers are deeper than freshwater bodies. The water from this source is sometimes referred to as brackish water.

Scale. This is a crusty deposit formed by compounds of calcium, magnesium, and barium, to name the principal species. Collectively these are classified as divalent ions. In common household usage their removal is known as water softening.

Shale. A rock formed by the deposition of sediment, usually transported by water, and comprising primarily clay and silt. This type of rock is usually deposited in characteristic layer patterns.

Shale gas. Natural gas found in shale bodies.

Shale oil. A mature form of oil in shale bodies, not to be confused with *oil shale*, which is shale that contains kerogen.

Slickwater fracturing. This term is used when the fracturing fluid used has substantially no gelling agent, such as sugars. Consequently neither crosslinkers nor breakers will be in use. Much of shale gas drilling uses this technique.

Source rock. A shale body with oil or gas in fully mature form, which usually is produced only after it has migrated to a porous body such as sandstone or carbonate.

Spur line. Short-distance pipeline connecting the producing rig site with a major gas export line.

Syngas. Short for synthesis gas. This is a mixture of carbon monoxide and hydrogen produced from any carbonaceous source, often natural gas. Syngas is the basic building block for a host of chemicals including methanol and ethers.

Thermal cracking. A refining process involving heat to break down molecules to smaller sizes.

Thermal maturity. This relates to the process of heat and time breaking down organic molecules to first form kerogen and then oil and gas. The stage at which any particular reservoir is in this process is described as the thermal maturity of the reservoir.

Tight gas. Natural gas found in substantially impermeable rock. Shale gas falls in this class, as does gas in sandstones and in carbonates. The key distinguishing feature is the difficulty of movement of fluids through the rock. Fracturing the rock induces artificial permeability.

Total organic content. Usually abbreviated to TOC. This is a measure of the economic potential of hydrocarbon-bearing shale bodies.

Wet gas. Natural gas with enough associated NGLs to materially improve the economics of recovery.

WTI. West Texas Intermediate. This is one of the benchmarks in oil pricing and comprises light sweet oil with an API gravity of about 39.6. The other important benchmark used is known as Brent.

Introduction

"Everybody look what's goin' down"

—From "For What It's Worth" by Buffalo Springfield (written by Steven Stills)

Shale gas is a newly discovered resource that could make the US self-sufficient in natural gas for over a hundred years. It has already enabled the US to overtake Russia as the largest producer in the world. It is uniquely enabled by a technology known as hydraulic fracturing, a technique under heavy scrutiny for the risk of environmental damage. Some seek to ban the method, and entire states and one country have issued moratoria.

The primary purpose of this book is to shed light on every issue that I consider germane to shale gas and the enabling technology, fracturing. My intent is for supporters and opponents to learn something they did not know about the issues before reading the book. Perhaps it will cause some to rethink their positions. I know that in researching the issues I became aware of some facets that surprised me. One was the positive impact of shale gas on national security. Another was an understanding of why inadvertent releases of methane occur during gas production.

Others will find ammunition to support their held beliefs. That is fine as well. At least the debate will be joined with factual support. A more informed debate is all one could hope for. This is a critically important issue for the nation, and the world by extension, and we must get this one right.

Responsible production of shale gas offers the promise of low-cost energy for a long time, and not just in the US. At a Department of Energy conference in February 2012, Bill Gates emphasized the connection between cheap energy and improvement of the human condition: "If you want to improve the livelihoods of the world's poorest 1 billion, their access to cheap energy determines if they can afford fertilizer, transport, and lighting—the things we take for granted as part of our lives and our dignity. Without advances in energy, they will remain stuck where they are" (as cited in Johnson, 2012).

In this context, telling is the fact that India imports natural gas at a price that in April 2012 was nearly *six times* the price in the comparatively affluent US. Domestic shale gas would make a dent in costly Indian imports. By learning to responsibly produce shale gas, the US is in a position to materially improve the human condition. This book takes a modest crack at discussing all the elements required to achieve this result.

The book is written for general consumption and so is necessarily short on technical depth. But underlying scientific rigor is certainly the intent. Calculations and chemistry details are relegated to shaded boxes, which can be skipped without losing the main thread. Also, throughout the book the term "gas" is used solely to mean natural gas. When gasoline is being discussed, the word is spelled out. The common American term *gas* is not used in that context.

The book approaches the topic from the standpoint of the economic value, the environmental hazards and potential remedies for them, and the political dimension. Balance is sought in the discussions, but ultimately balance is in the eye of the reader. If I have a bias it is toward seeing problems as opportunities to devise solutions and create economic value. Other than that, I have the bias of most people, of ensuring access to affordable energy while minimizing environmental risk.

The discussion is primarily US-centric because the intended audience is largely domestic. However, the principles apply elsewhere, and where relevant those implications are discussed, such as in chapter 22, "Kicking Shale Into the Eyes of the Russian Bear," which also discusses the use of gas supply as a political tool. Much of the world is watching to see whether we can exploit this economic windfall with minimal negative impact on our environment and way of life.

Much of the book appears to have a focus on the present. In part this is deliberate. The realizations regarding the size of the resource, the impact on the economy, and the hazards associated with potentially careless production are relatively recent. Public anxiety is at a higher level than it was after the *Deepwater Horizon* oil spill five years ago. Decisions made today will affect our economic and environmental security and well-being for decades. So a focus on the present is valid. However, the book also contains discussions about the future—in particular, about the predictions of the price of natural gas and oil, which have significant bearing on decisions in the coming years. The future is also discussed in the context of alternative fuels, which necessarily take multiple years to have material impact.

Environmental activism has been successful in closing coal-based electricity-generating plants, particularly older ones. This capacity cannot be replaced by renewable sources such as wind and solar in the short term because they are still generally too costly. I subscribe to the view that natural gas ought to be used as a bridging fuel until renewables are economically viable for base load. However, in chapter 21, "Will Cheap Natural Gas Hurt Renewables?" I discuss the belief that low gas prices will hurt the cause of renewables. This could change with price rises caused by increases in demand. The circumstances that could cause increases in demand are discussed in several chapters.

The book is laid out in sections, with fairly self-contained short chapters on each topic. The intent is to allow the reader to skip to any chapter with a reasonable assurance of comprehension of the issues presented. But inevitably the understanding of some detail may well require that another chapter be read; in these cases references are made to the chapters in question. In response to an anonymous reader's critique on Amazon.com, we added a "big picture" introduction at the start of each section.

While there is a solutions-oriented flavor to the environmental issues throughout, there is little doubt that this will be a tough row to hoe. It will need a combination of legislative oversight, technical innovation, and industry doing the right thing, with the last being nudged along by informed activism in the communities in which they (will) operate. The triple-bottom-line tenets of sustainable energy production must be put into real practice. In plain words: without profit there is no enterprise, but it must not be at the expense of either the obligation to environmental stewardship or the interests of the local community.

PART I

Shale Gas Basics

I. Shale Gas Basics

Any book dealing with the comprehensive aspects of oil and gas production needs to start by explaining the basics—how oil and gas are formed and where they are found. This is even more important when talking about shale oil and gas because this type of resource is not well understood even by many in the industry. As a result, production methods are not very sophisticated and only small fractions of the oil and gas in place are successfully produced at this point.

The general public ought to be even more bemused by the notion of massive resources when none were previously known to exist. When one examines the precise mechanism of formation of oil and gas, it soon becomes apparent that the resource did not appear out of nowhere. It was always there but not deemed accessible. The new technologies of horizontal drilling and hydraulic fracturing made it possible to tap into these resources in surprisingly prolific ways. However, these technologies also brought environmental baggage, the sole subject of Section II.

Shale gas could eventually displace coal for electricity generation, and here I demonstrate the relative economics using the two different fuels. In my opinion, natural gas prices will remain competitive with coal for decades. We can look to a future of steady coal replacement.

The oil price crash of 2014/2015 changed the world order in oil and gas. Shale oil has shifted the balance of pricing power away from OPEC to the US. I explore oil market economics in this section.

CHAPTER 1

So, Where Did All This Gas Come From Suddenly?

"Do you believe in magic?"

—From "Do You Believe in Magic" by The Lovin' Spoonful (written by John Sebastian)

Few will dispute that shale gas has changed the very makeup of the petroleum industry. At every twist and turn new resource estimates appear, each vastly greater than the previous. The estimate in 2008 exceeded the one from 2006 by 38 percent. Each year the US Energy Information Administration (EIA) estimate of the resource is higher than the previous. This is to be expected for a new type of resource: commercial activity unveils new deposits. The 2013 estimate is 10 percent greater than the one from 2011. As with all resource estimates, be they for rare earth metals or gas, disputes abound. But through all the murk is the inescapable fact: there certainly is a lot of the stuff. How could this suddenly be so? The last such momentous fossil fuel find in North America was the discovery of Alaskan oil. But a discovery out in the far reaches of the US is understandable. In the case of shale gas, all this is happening literally in our backyard.

To appreciate the excitement at the discovery of shale gas we first need to understand how oil and gas are formed and recovered. Millions of years ago, marine organisms perished in layers of sediment comprising largely silt and clay. Over time, additional layers were deposited and the organic matter comprising the animals and vegetation was subjected to heat and pressure. This converted the matter into immature oil, known as kerogen.

Further burial and temperature rise continued the transformation of kerogen to oil. The most thermally mature final form is methane, formed by the thermal cracking of oil. By and large the only real difference between oil and gas is the size of the molecule. Methane is the smallest gas, with just one carbon atom. One of the lightest oil components, gasoline, averages about eight carbon atoms. Diesel averages about 12. So, although we refer to them as oil and gas, chemically they are part of a continuum, so it is easy to understand that they come from a single source.

Oil Shale

Kerogen left in the original immature form is also found in certain shale deposits. These shale deposits are known as *oil shale*. The nomenclature can be confusing because a mere switching of the two words leads to a completely different product. (The oil found in shale is known as *shale oil* and is light oil in mature form ready to be distilled into useful products such as gasoline.)

Oil shale resources are estimated to be very large, greater than all conventional oil combined. The US has the largest deposits, primarily located on federal lands in Colorado, Utah, and Wyoming. However, the difficulty of extraction and processing has made these resources largely academic except in a few countries. Estonia is by far the most prominent user of oil shale, using it for electricity generation.

The immature kerogen requires processing to simulate what millions of years of heat and pressure accomplished in nature. Many of the deposits are very shallow and so can be mined. The mined material can be thermally treated to produce a hydrocarbon vapor, which when condensed yields a useful oil. By and large this is considered too expensive today.

Royal Dutch Shell conducted considerable research into heating the kerogen in place using a pattern of resistance heating elements. This process was piloted in Colorado but eventually mothballed. In principle, this in situ conversion is an elegant solution and likely the way forward in the future.

Petroleum

Petroleum principally comprises long-chain molecules with the formula C_nH_{2n+2}. In its simplest form $n = 1$, and this is methane, which is a gas at normal temperatures. When $n = 2$, this is ethane, also a gas, and one which features in later chapters. Propane has $n = 3$ and is a liquid at moderate pressure. Propane is most familiar as a cooking and heating fluid. Crude oil generally comprises long-chain molecules with n greater than about 20.

The greater the number n, the heavier the oil, because longer chain molecules are more dense. Refining is the process of breaking the molecules down (*cracking*) and often adding hydrogen as well, known as *hydrothermal cracking*. When all the oil produced was pretty light, with low n's, all one had to do was to heat it up. The different molecules would be separated based on their condensation temperature. This is known as *fractional distillation*.

The key word is *source*. The rock in which the oil or gas (or both) originally formed is known as source rock. In Figure 1, the source rock is gas-rich shale. The source rock is almost always shale, which is some mixture of silt and clay and sometimes some carbonates. Conventionally, the fluid in this rock migrates to a more porous body.

Figure 1. Schematic of the geology of natural gas resources

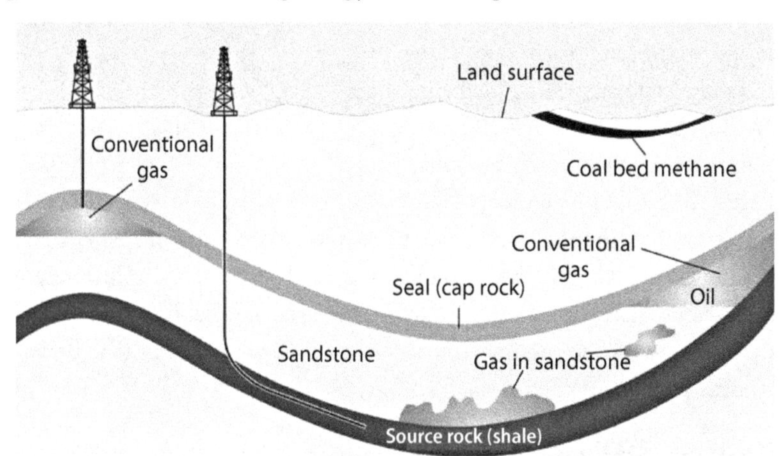

Source: US Energy Information Administration, January 27, 2010

The porous body to which the fluid migrates is depicted in Figure 1 as sandstone, which is predominantly silica, an oxide of silicon. It may also be a carbonate, predominantly calcium carbonate. These two minerals are host to just about every conventional reservoir fluid in the world.

The fluid (and by the way, gas is a fluid, although not a liquid) migrates "updip" (up the slope of the rock formation), as shown at the upper left and right of Figure 1. This is because the hydrocarbon is less dense than the water-saturated rock and essentially floats up, not unlike the oily sheen on your cup of coffee. This migration continues until it is stopped by a layer of rock through which fluid does not easily permeate. This is known as a *seal*, or more colloquially, a cap rock. Ironically, this is most usually a shale, not unlike where the fluid (gas) originated. The trapped fluid is then tapped for production.

The trap is often in the shape of a dome, as shown in the upper left. It can also be a fault. This is when earth movements cause a portion of the formation to break away and either rise or fall relative to the mating part from which it just separated. In some instances a porous fluid-filled rock will butt up against an impermeable one, and a seal is formed laterally.

In Figure 2, the lightly shaded zone represents sandstone, and the fluid of interest, shown in the triangular hashed area, abuts the cap rock, an impermeable zone shown in dark gray. This sort of faulting happened a long time ago, so the fault has healed. There is no fluid escape path along the fault.

Figure 2. Schematic of a fault with trapped fluid

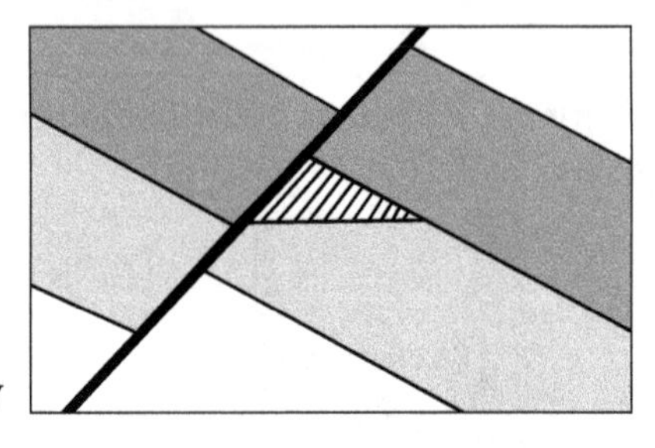

In the early days of prospecting, explorers looked for surface topography indicative of a dome-type trap below. Nowadays sound waves are sent down into the formation, using small explosions in most cases. The sound waves are reflected back, and an analysis of the various reflections produces excellent images of the subsurface. This is fairly similar to ultrasound imaging of the internal organs of a human body for diagnostic purposes.

Unconventional Gas

I have described how conventional gas—and oil, for that matter—are found and produced. The current flurry of activity in shale gas is concerned with going directly to the source. This was previously considered impractical, primarily because the rock has very poor permeability, which is the ease with which fluid will flow in the rock. The permeability of shale is about a million times worse than that of conventional gas reservoir rock. In fact, as I observed earlier, shale acts as a seal for conventional reservoirs. The breakthrough was the use of hydraulic fracturing. In hydraulic fracturing, water is pumped below ground at high pressure, causing a system of fractures in the rock into which the fluid is injected. These fractures are then propped open with some ceramic material, most commonly sand. Without this the sheer weight of the thousands of feet of rock above would close the cracks. The propped open fractures now constitute a network of artificially induced permeability, allowing the gas to flow out of the formation and up the borehole to be collected (in industry parlance, to be "produced"). This is akin to the use of pillars and beams in underground coal mines, in that case allowing passage of people and bins hauling coal.

The sheer ability to extract gas from source rock is now well understood as feasible. But some still doubt the magnitude of the estimated resource. Here is the explanation of why one would expect this resource to be plentiful. Consider that for a conventional reservoir to be formed, two events had to occur. First, there needed to be a proximal porous and permeable rock, and second, a trap mechanism had to exist. It would be easy to believe that more source rock did not have these conditions than did. In other words, the probability that fluid in source rock does not have a way to permeate to more porous rock surrounding it is greater than the probability that it does have such a release mechanism. This is why it is a reasonable conjecture that the total oil or gas trapped in source rock is greater than the amount of it that escaped into permeable trapped rock such as sandstone. Further adding to the potential is that this is fresh territory, relatively unexploited. Decades of exploitation have denuded conventional reserves, while the source rock further below remains relatively untapped.

A word on the nomenclature of resource estimation. A *resource* estimate indicates the quantity of estimated hydrocarbon accumulation, whether economically recoverable or not. A subset of that is a *reserves* estimate. Reserves are the portion of the resource that one could recover economically and bring to market. Typically in a new play (the term oil folks use for an active prospect) one would expect reserves to keep getting revised upward. This is because every new well put on production increases the certainty of the extent and quality of the reservoir, and the reserves can confidently be increased. In reading the popular literature it would be well to keep the distinctions in mind; they are often confused.

All of the press has been about North American activity. But once the basics are understood one can readily believe that source rock is ubiquitous. The EIA published a report (EIA, 2011, April 5) in mid 2011 providing its estimates of resources worldwide. China was seen with the most, followed by the US and then Argentina. The 2013 update (EIA, 2013, June 13) left China in the lead, with Argentina number two and Algeria slipping in ahead of the US into third position. This report also for the first time estimates shale oil, with Russia in front. More to the point, EIA estimates that source rock occurs worldwide, as I have conjectured. The greatest impact will be on countries that are net importers, such as Poland, South Africa, Turkey, and the Ukraine. Poland is already aggressively pursuing development of the resource base, but with limited success to date. The have-nots of the hydrocarbon world can kick up their heels now.

CHAPTER 2

The Oil Plateau and the Precipice Beyond

"Up, up and away"

—From "Up, Up and Away" by The 5th Dimension (written by Jimmy Webb)

Christophe de Margerie, the CEO of Total, based in France, made a startling pronouncement in the fall of 2007. He said that oil production would soon plateau, at the level of 100 million barrels per day (MM bpd). Needless to say, it made quite a splash. Here was the CEO of one of the top five oil companies in the world saying there's a plateau coming. At that time the world was producing about 85 MM bpd and consuming the same amount. In fact, that has been the pattern: consumption equals production, and price modulates demand.

After that I personally, publicly asked a CEO of a major oil company to comment on de Margerie's prediction. He acknowledged the plateau as being real. He said, "I'm not sure I'm going to subscribe to the 100 number, but there's a plateau coming." Shortly before that I spoke to the head of the French Petroleum Institute (IFP), who confirmed that their modeling showed the same thing. They pegged it at a somewhat lower number.

So here we have substantial people saying there's a plateau coming and yet nobody acknowledges it publicly. Nobody wants to discuss it. Nobody really wants to act on it.

The significance is that if this is supportable, and a case can be made for recovery-driven economic growth, a supply/demand imbalance will cause havoc. The solution in part is to find substitutes for oil. Cheap natural gas can be one avenue, hence the inclusion of the oil plateau discussion in this volume. But first the thesis.

Causes

Now you'll ask the reasons for the plateau. First of all there is a technical model that predicts a plateau, courtesy PFC Energy in Washington, DC (PFC Energy, 2009), which I discuss to some degree below. But some qualitative arguments can be made in support of a plateau hypothesis. For example, national oil companies have realized they have a resource they need to husband. International oil companies used to move into resource-rich countries and extract oil via production-sharing contracts. Such contracts had built in the incentive to get the most oil out as quickly as possible in part because the contracts had limited duration.

There's a truism in oil and gas production: if you draw it out quickly, then the net recovery—that is, the fraction of fluid in the reservoir that is ever recovered—reduces. The industry currently leaves behind about two-thirds of the oil in place. When the international oil companies went into these resource-rich nations, they drew oil out as quickly as they could because that maximized the value of the contracts. That was not in the best interest of the national resource.

Increasingly the nations with oil resources have figured that out. Now they are forcing the issue, telling the international oil companies, "We'll do it ourselves. We don't need you." The key point is they want to bleed the oil out in more measured fashion. Guess what that does to production rates?

Most of the major oil companies are therefore forced to seek unconventional sources of oil—for example, Canada's tar sands—which are largely heavy oil. Additionally, Alberta now has a small carbon tax on oil from the tar sands.

The late Matt Simmons, a highly respected figure in oil and gas investment circles, wrote that Saudi Arabia would not be able to open the spigots—that they didn't have the oil (Simmons, 2005).

Predictably, the Saudis have been mum on the point. The fact of the matter probably is that the Saudis have the oil, but they've got a different view of it now and how to release it. The national oil company Aramco has been a leader in the application of technologies to maximize recoveries and so can be expected to add to the nation's productive capability. But they're not going to get bullied into releasing it faster just because the world wants a lower price of oil. People thought of Saudi Arabia as the buffer, that they'd just open the dams, but it just doesn't seem like they will. Matt Simmons took the position that they could not. It's irrelevant: they won't. Whether they can't or won't, the result is the same: they will no longer make up shortfalls elsewhere unless driven by political exigency.

Consumption versus Production

The estimated production plateau of 95 MM bpd—I think PFC at this point is talking about 90–92 MM bpd—comes dangerously close to the 87 MM bpd we were consuming prior to the recession. In mid-2015 that consumption (demand) is still down around 85 MM bpd. Following economic recovery, consumption had risen to 89 MM bpd by the end of 2012, but most of this came from non-OECD (Organisation for Economic Co-operation and Development) countries. US consumption continues to drop year after year, as does Europe in aggregate. The big increases are in Asia, and China in particular, as would have been predicted by per capita GDP growth. China is in the early stages of addressing this through methanol (from coal) substitution of gasoline.

Figure 3 is derived from data from PFC Energy. Their demand curve crosses over the supply around 2020. This is a scant five years away. PFC makes the point, as do I, that the only real solution is finding alternatives to oil utilization. Note that they show the plateau for only few years, and then a decline. All of this is subject to speculation. But the key takeaways are as follows: more oil has to be found and produced, and alternatives to oil urgently need to be developed. The urgency draws from the fact that capacity ramp-up of biofuels and the like takes billions of dollars and multiple years.

Figure 3. World oil supply versus demand projection

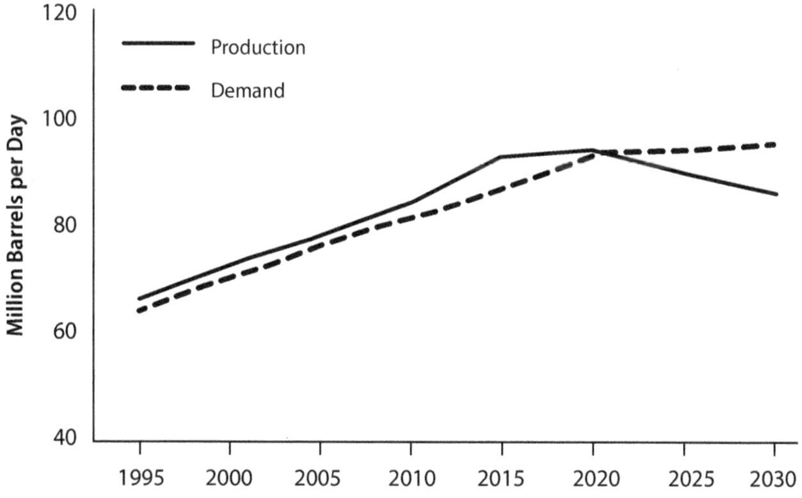

Production
Demand
Million Barrels per Day
120
100
80
60
40
1995 2000 2005 2010 2015 2020 2025 2030

Source: Adapted from PFC Energy data (PFC Energy, 2009)

The key factor is the speed of the recovery with respect to automotive use. In the United States, at least, oil is about transportation. Gas is about power and petrochemicals. The plateau is real and the recovery is real. It's very real in China and India, which never really saw much of a recession. They each have had close to double-digit growth in GDP for the last several years and continue apace. In China and India, what do you think newly prosperous people do? They buy a vehicle. They go from a bicycle to a motorcycle or scooter to a car.

Shown in Figure 4 is per capita car ownership on the vertical axis as a function of per capita GDP. China is near the bottom, and India is even further below, almost off the graph. These are countries with the fastest growth. The US is on the upper right, the direction in which these others are headed.

Figure 4. Per capita car ownership as a function of per capita gross domestic product (GDP)

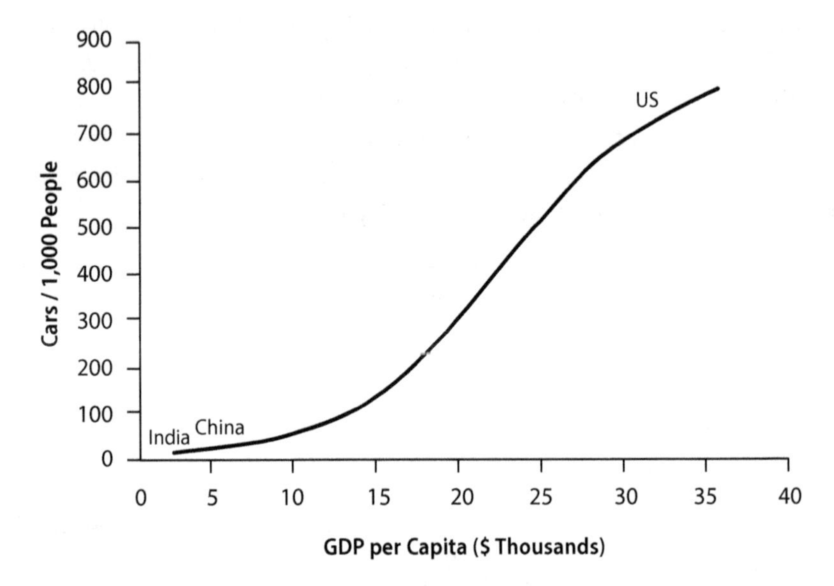

Source: Adapted from PFC Energy data (PFC Energy, 2009)

All of this indicates that transport fuel usage is likely to keep increasing, and that if it does, the crossover point between consumption and production is probably sooner than later (I'm not talking electricity—that's a completely different argument).

Any target for reduction in imported oil requires displacement of oil-based transport fuels. The rise of wind- and solar-based electricity cannot be juxtaposed with oil usage reduction. This is because very little transport runs on electricity. Years from now, when electric vehicles are a significant fraction of active automobiles, that could change.

The plateau is coming, and if consumption continues at the current rate, a crossover is coming, too. And at the point of the crossover, we're not talking a spike in prices. We're talking a sustained price increase. A spike is driven by a shortage at some point. This is not a shortage at some point. This is a plateau. Clearly demand destruction will occur in response to this. But the initial surge in price will be damaging.

Here is the crux, though: Do you really want to test the plateau theory? The alternative to testing it is doing something smart, like displacing oil with something that is more environmentally responsible.

The Role of Shale Gas

Prior to the intruding reality of massive amounts of cheap natural gas, the options for oil substitution were limited. Electrification of transportation was, and remains, the most important avenue. However, by its very nature, the carbon footprint burden is merely shifted elsewhere. The electricity generation process generates carbon dioxide, even though the vehicle emits none. But certainly oil would be saved. The Fukushima Daiichi nuclear disaster placed a damper on nuclear energy as a source of clean electricity. Germany placed a moratorium on new plants, as did Switzerland, and the raising of the risk bar has caused funding to dry up in other countries. That pretty much left coal, natural gas, and renewables.

Clean coal is simply not going to have legs until we have a price on carbon, be it a tax or some other measurable way to pay for cleaning it up. Is it technically feasible? Yes. But it will add in the vicinity of 3–5 cents per kWh to clean it to natural gas levels. Aging plants will be loath to add new equipment no matter what the incentive.

Wind and solar will continue to get increasingly viable. However, they simply will not make a major dent for decades. Also, to be viable alternatives, they need a price on carbon (in the form of a tax or penalty associated with carbon emissions) or policy boosts. The solar subsidy in Germany appeared to work, but the post-subsidy world will be rough.

That leaves natural gas. Shale gas is causing the price to remain low in North America. There is reason to believe that similar effects will be felt elsewhere. A dramatic shift from coal to natural gas for electricity is likely and feasible.

Shale Oil and Natural Gas Liquids

This is a huge wild card in the oil game. First some definitions are in order. Shale oil is oil found in shale, much as the gas is. This is distinct from oil shale, which is sedimentary rock containing an immature form of oil that needs significant processing. It is generally believed that the oil shale resource, also vast, is not feasible for exploitation in the foreseeable future. North America is especially blessed in this regard, but technical breakthroughs are needed to economically "cook" it in place to make its use viable.

Shale oil is found in two settings. One is in the oil portions of shale gas reservoirs, and the other is in completely different areas with varying proportions of associated gas. The outstanding example of that is the Bakken formation in North Dakota, Wyoming, and Montana and extending into Canada. As in the case of shale gas, fracturing is necessary to release the liquid. The production rates are modest per well. But the production is economical and the resource estimates are huge. Unlike that other unconventional oil, heavy oil, shale oil is light and sweet, meaning it is low in sulfur, and the size of the molecules is on average small. These characteristics by and large ought to make it a refinery darling because it is less costly to convert to gasoline. However, US refineries do not welcome this liquid. This is discussed in chapter 12, but the abbreviated reasoning is that the refineries have expensive equipment designed to treat cheap heavy oil. Decreasing utilization of the equipment while paying a higher price is not something they want to do.

Assuming production from shale oil is substantial, how does one rationalize the existence of an important resource against the plateau effect? For one thing, the plateau is likely driven by lack of access to the resource base. The US, as a net importing nation and one not limiting access to at least this type of resource, is positioned to become an important new producer. The US consumes about a fourth of all oil produced, and significant domestic oil production will eventually affect world oil prices. But that is not on the medium-term horizon. The very nature of the shale oil resource limits the growth ramp. But it could materially affect the supply/demand equation.

Natural gas liquids (NGLs) are liquids found in association with shale gas. Because the price of gas is low, and will likely remain low for years, most of the activity is likely to be in the portions containing NGLs, known as wet gas. These are portions of the reservoir with a higher proportion of the liquids. The liquid price is pegged to oil prices, not gas. If oil remains tight in supply, as hypothesized above, the price can be expected to remain high. Today, in early 2015, oil is roughly three times the price of gas on the basis of energy content. One can expect that ratio not to get better and possibly to get much worse in the coming years. So wet gas is where the action is and will stay until gas prices rise. That could happen with demand improvement. But wet gas will always be more profitable.

Wet gas will have liquids associated with natural gas in quantities ranging from 4 to 12 gallons per thousand cubic feet (mcf). In the Marcellus Shale, it averages 7 gallons per mcf. That is 0.17 barrels, since each barrel is defined as 42 US gallons. At today's price in early 2015 of $65 per barrel, the value of the liquids will become commensurately less. For natural gas we will assume a price of $3 per mcf, even though it has seen spot price drops during warm winters and rises of over a dollar in cold ones. The liquid component is worth 0.17 × $65 × a discount factor added for conservatism. Even taking the discount parameter to be 0.3, the liquids are worth $3.30, while the associated natural gas is worth $3. No matter what reasonable discount you apply, the liquids materially add to the profitability of the gas. Small wonder that wet gas is the play.

In chapter 14, I discuss the ethane dilemma. One reason the discount factor is low is that ethane, which can constitute over half the liquid (actually ethane is a gas in that state, but it is lumped in with the NGL by the industry), is priced half that of the bigger molecules such as propane and butane at this time. Ways to monetize this are discussed in that chapter.

CHAPTER 3

Gas Will Remain Cheap and Displace Coal

"She got the goldmine, I got the shaft"

—From "She Got the Goldmine" by Jerry Reed (written by Tim DuBois)

Natural gas will rapidly displace coal for power even without the benefit of a price on carbon.

When natural gas is combusted for power production, the carbon dioxide produced is about half that produced from coal. Beyond this, the externalities associated are small when compared to those attributed to coal consumption, notably the environmentally important (and with health implications) possibility of emissions of mercury, sulfur, NOx, and particulates, and less noticeably fly ash disposal problems. A decided shift toward natural gas in the early years of this century stalled when gas prices became unpredictable. The shift started when gas was at about $2 per million Btu (MM Btu) and the all-in cost of electricity production was less than with coal. A Bernstein report (Wynne et al., 2010) details the history of this, including the quandary reached in 2007 when gas prices were going up but so were construction costs for coal plants. This caused a large number of planned coal plants to be postponed.

Today Congress has reached no consensus on legislation to mitigate carbon emissions, and the political makeup of the current Congress makes that prospect less likely. So, a price on carbon is unlikely in the near future. Mind you, the European version of cap and trade has simply not worked. For years the price has fluctuated from about €15 ($19.50) per metric ton (tonne) to about €25 ($32.50). In late December 2011 it actually crashed to €6.50. Uncertainty in price dampens investment because there is no choice but to make the discount rates higher. That raises the overall cost of investment.

However, the US Environmental Protection Agency (EPA) is expected to levy strong restrictions on mercury, sulfur, and NOx. Their hand has been strengthened by the April 2014 decision by the US Supreme Court, which endorsed the EPA's Cross-State Air Pollution Rule that seeks to limit power plant emissions in 28 upwind Eastern and Midwestern states to help reduce air pollution levels in downwind states. One report (Wynne et al., 2010) states that more than 40 percent of coal generation plants that do not meet the latest EPA standards are over 50 years old. One can reasonably expect many of these to be mothballed, leaving room for new coal plants, gas-fired plants, or alternative methods such as nuclear, with the problem exacerbated as the global economy continues to expand. With gas plants costing less to build, being cheaper to operate (at forecast gas prices), and easier to obtain regulatory approvals for new construction, the bias toward gas power generation will be significant. The Fukushima Daiichi nuclear plant disaster in Japan has had a chilling effect on the nuclear option. Switzerland banned the building of new nuclear plant reactors, and Germany went back on its agreement to extend the life of existing plants. Some of this action occurred within days of the tsunami and subsequent reactor meltdowns and there was no firm finding regarding the true impact. Nobody is seriously considering wind or solar for base load any time soon.

The lead time to first electricity production is low for gas, as compared with coal or nuclear. Coal and gas are polar opposites on components of cost. For coal the capital cost is a high proportion of the all-in cost, around 60 percent, whereas the commodity is relatively stable. In the year 2010, coal varied little from a cost of about $2.25 per MM Btu. The latest EIA forecasts suggest similar stable behavior: averages in the vicinity of the 2013 price of $2.35 per MM Btu out until 2016. This is the best way to express the cost of coal because the Btu content is highly variable; from brown coal to anthracite, the Btu content differs by more than a factor of 2. So in comparing coal with other fuels, the price per ton is not a meaningful figure, though oft quoted in reports. Btu content is a better measure of comparison.

Gas-based power, on the other hand, has a relatively low capital cost, around 15 percent of the all-in cost, and in the past a highly variable price, at least over the lifetime of the amortization of equipment. Consequently, while coal power is highly susceptible to construction costs and inflation in general, gas economics depend primarily upon the price of the commodity.

Figure 5 plots the all-in cost for electricity produced from natural gas as a function of gas price. The plot uses data from the Bernstein research report cited above (Wynne et al., 2010) and reflects 2006 economics. The price in October 2010 is shown in the figure. It dropped under $3.50 in 2012 and was episodically over $5 per MM Btu during 2013 and also below $4 at times in 2014. The generally accepted cost of production from new coal-based construction meeting emission standards is drawn as a horizontal band between 6 and 6.5 cents.

The impact of a price on carbon can be treated in one of two ways. Post-combustion carbon capture using technology known today, but not yet perfected, can be expected to cost about 3 to 5 cents per kWh to bring the carbon dioxide emissions down to natural gas carbon emission levels. This is plotted as the second horizontal band above the first, plotted using the more conservative figure of 3 cents. The second way to treat the impact of a price on carbon is to simply pay the carbon penalty. At a price of $25 per tonne this will add around 1 cent. This accentuates the point that any price less than about $40 per tonne is simply not reflective of the cost of taking the carbon out.

Figure 5. Comparison of electricity cost: coal versus natural gas, as a function of gas price

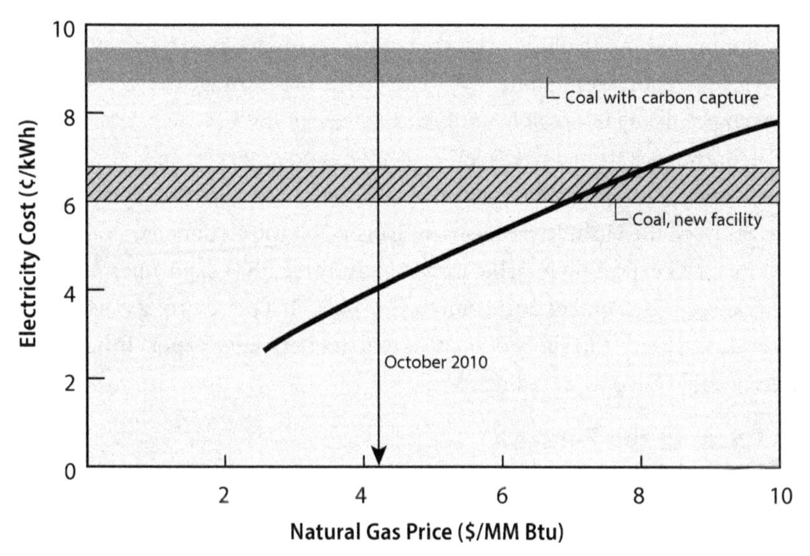

Source: Wynne, Broquin, & Singh, 2010

Of immediate note is the fact that in 2013 the gas-based cost was well below that from a new coal plant. Of particular note is the observation that the breakeven with coal kicks in at a natural gas price between $7 and $8 per MM Btu. This is without consideration for any carbon penalty. If that were to be included, the breakeven moves to gas prices between about $9.50 and $12 per MM Btu, depending on which model of carbon penalty one uses. The gas price spikes of several years ago endure in memory. The fact is that prices were over $12 for only about four noncontiguous months. Predictability of price and assurance of supply are needed for a major shift from coal to gas.

Assurance of Supply

A scant three years ago this issue would have had a different answer. We would have been discussing liquefied natural gas (LNG) imports and environmental risk posed by that. Shale gas has changed all of that. The US can expect to be self-sufficient for a hundred years. In fact, it may well become a net exporter of gas or gas derivatives.

This could also change the entire debate around natural gas in Prudhoe Bay, Alaska. Natural gas in copious quantity has been reinjected (about 8 billion cubic feet [bcf] per day) because of the high cost of a pipeline to the Lower 48. Under the circumstances a substantial bullet got dodged by the delay in that decision. Had a pipeline been built, the fully loaded cost of the gas would have been challenged by that of shale gas production and the utilization could have plummeted. As an alternative, the nation should reopen the possibility of exporting Alaskan North Slope LNG. Due to the net shortage in the US this had been politically impossible until recently except for a single permit for LNG export of gas from Cook Inlet. That should no longer be an issue, and in fact ConocoPhillips has received an extension to the Cook Inlet permit. Shale gas from the Horn River Basin in British Columbia (Canada) is already slated for LNG export by Apache. Importantly, three LNG export permits were granted by the end of 2013, amounting to 5.5 bcf per day of gas usage. Expect this to double in short order. Gas and gas derivative export from North America could become a trading force.

Gas Prices in the Future

The floor will be determined by demand. In 2013 it hovered around $4 per MM Btu with only modest demand creation. A continuing shift from coal to gas will drive demand. So the concern is likely more at the upper end. As I have discussed elsewhere, assurance of a moderate price is a huge driver in the chemical industry. Is there a mechanism for a ceiling? The answer is affirmative and results uniquely from the setting in which shale gas is found.

Most shale gas is either proximal to the intended market, as in the case of the Marcellus, or close to major pipelines, as in the case of the Barnett (Texas), Haynesville (Louisiana), and Woodford (Oklahoma) shales, to name just three big ones.

Compared to conventional gas, these wells are relatively shallow and on land. In a fairly recent study, examination of 100,000 wells showed that the average time for the construction of a horizontal producing well from the beginning of a well to the start of the next one was 27 days (Nikhanj & Jamal, 2010). With the steep learning curve to which this industry is accustomed, these numbers improved substantially in 2014. In fact, shale gas is particularly advantaged in this respect because the sheer number of wells allows for innovation to be perfected in what amounts to a factory situation. However, the time from the start of drilling to the point of delivery to sales has been much greater, averaging closer to 150 days. This does vary by area. The Haynesville Shale, with the longest drilling times due to its depth and complexity, curiously has the shortest time to market, about 90 days.

Pad Drilling

The concept of drilling up to 15 or more wells from a single location (pad) was a technique pioneered in Colorado as an environmentally friendly technique to minimize roads, among other features. However, this may be the culprit for the long time to market, because it allows wells to be drilled and completed in batches. Batch drilling is a technique wherein the early portions of a number of wells are completed at the same time, followed by the later portion of the wells all together. This reduces overall cost in many instances. But it necessarily greatly increases the time from first well start to first fluid delivery.

That said, the pad technique is likely here to stay, particularly in the heavily populated areas of the Marcellus. It dramatically reduces the traffic associated with pump trucks, proppant delivery, water disposal, and the like. The aggregation also permits higher levels of sophistication in areas such as water handling, remote decision making, and quality control. Ultimately operators will balance the economics of batch operations against the opportunity cost of later sales.

So, depending upon the area and the business drivers, new production can be brought to bear in as little as 90 days and certainly under 180 days. This is compared to over four years for a conventional offshore gas field. This short time span will basically keep a lid on the high end. Speculators will be aware of the quick response ability. This ability to service demand by bringing on capacity at short notice will likely keep prices under $8 per MM Btu, possibly closer to $6 per MM Btu. This is the crux of the thesis that shale gas will enable natural gas to be in a tight band between $4 and $6.50 per MM Btu for a long time.

At numbers north of $4 per MM Btu most operators will make a very good profit, especially if a portion of the gas is laden with natural gas liquids. Newer technologies will continue to drive down cost, as has been the case in every new resource play since Colonel Edwin L. Drake's historic oil find in Pennsylvania in 1859. So the businesses will be sustainable and continuous supply assured. Note: LNG has a built-in cost of nearly $3 to $4 per MM Btu just to liquefy, transport, and re-gas, over and above the gas production cost.

In summary, shale gas has changed the game for all industry and power in particular. For the first time in memory, a major fuel can be predicted to be priced in a tight band at low to moderate levels for many years. This is the type of certainty that drives investment. The timing of the realization is propitious in dealing with upcoming new EPA regulations on coal emissions and possible regulations on CO_2. No new coal generation plants are economically justifiable provided our predictions on natural gas pricing hold up.

CHAPTER 4

What a Difference a Hundred Million Years Makes

"Time is on my side"

—From "Time Is on My Side" by The Rolling Stones (written by Jerry Ragovoy)

Utica, New York, is an unpretentious city of 60,000 people that has an interesting geological feature. Here exists a rock outcrop which is named after the city: the Utica Shale. This shale body is just beginning to make waves in the natural gas industry. This rock is old—100 million years older than the Marcellus Shale, the current darling of the shale gas industry. The operative word is *current*. The Utica appears set to upstage the Marcellus.

The Utica Shale is directly below and greater in lateral extent than the Marcellus. If all of it proves productive, it will likely be the largest gas field in the world. If the initial production numbers prove typical, the Utica may be one of the largest in reserves as well. With the Marcellus itself no slouch, the two together represent an immense concentration of hydrocarbons. The Utica Shale extends into the Great Lakes. This offers the possibility of offshore shale gas operations, which would be a first.

The Utica is generally thicker than the Marcellus. In the primary regions of interest it appears to vary from 150 to 500 feet. Somewhat complicating the issue is that in some areas the fluid had migrated to sandstones and carbonates immediately above and adjacent to the shale, so the source rock may not be the only producer, unlike the Marcellus. The percent total organic content (TOC) is the amount of organic matter present and is a measure of the economic potential of the resource. This is somewhat lower in the Utica than in the Marcellus.

This difference in TOC is one of the reasons for caution regarding *resource* estimates as compared to *reserve* numbers. The latter value is always smaller for any prospect because to be qualified for a reserve designation, the fluid has to be economically recoverable and deliverable. In general, a TOC less than 2 percent is probably below the threshold of interest.

Productivity Considerations

Once the TOC threshold is crossed, the next consideration is the estimate of the gas in place. In conventional reservoirs, the gas is all free gas in the pore spaces. It is then driven out by a variety of drive mechanisms. Shale layers have two types of gas. The first is free gas in the pores. The other is gas adsorbed onto the organic matter. *Adsorption* is a surface phenomenon, as opposed to *absorption*, which is a bulk effect. An everyday example of adsorption is the activated carbon filter used in water filter systems to remove select harmful species such as bacteria. Activation is a process that dramatically increases the surface area of the carbon and increases its adsorptive ability. In the case of a water filter, the harmful species stay adsorbed on the carbon and the filter element is replaced from time to time.

In the case of shale gas, the adsorbed gas is released simply by reducing the pressure by opening up the well to the surface, and further reducing it through removal of the free gas. In principle one could improve desorption (the release of adsorbed gas) through injection of carbon dioxide. This molecule preferentially adsorbs on shale organic matter by about a factor of 5. Thus all the adsorbed methane would get released; this would be a CO_2 sequestration mechanism. The practicality of conducting this operation needs to be researched.

Shale usually comprises a combination of silica and clay, often with another mineral such as calcite or calcium carbonate. The relative quantities determine the ability to produce and propagate fractures. The best rock for successful fracturing is slightly brittle. Too much clay makes it too ductile. Think peanut butter as opposed to peanut brittle. See chapter 28 for another discussion on this topic with respect to Argentina's Vaca Muerta prospect.

What Makes Utica Different

The source rock belongs to the Ordovician Period, which extends to 500 million years ago and which is about as old as rocks derived from sediments can be. As a frame of reference, dinosaurs started 250 million years ago and got erased 65 million years ago. The black shale of the Utica is about 460 million years old, as compared to the Marcellus, which is around 370 million years old. This extra age manifests in two ways. One is that it is deeper wherever they overlay each other. This difference is up to 7,000 feet. The extra depth brings with it higher natural pressures for fluid production, which would be more similar to the prolific Haynesville Shale in western Louisiana and eastern Texas. Of course the deeper wells will be more expensive.

The extra age also brings with it a greater thermal maturity. This means a greater conversion of kerogen to oil and gas. All of this is good for productivity. In addition, the close proximity to sandstone and carbonate layers causes some carbonate content in the shale, making it more porous. This offsets the lesser TOCs compared to the Marcellus. But the carbonate content makes the formation more like the Eagle Ford Shale in Texas than the Marcellus in its fracturing character. So, despite the fact that Marcellus property leasers have the Utica below them, the techniques used would need Eagle Ford–type experience.

Producing the Utica

A point being debated is whether an operator could exploit both reservoirs at the same time. On the plus side, the drilling and fluid export infrastructure certainly could be used for both. On the other hand, with the Utica being at much higher pressures, it would be impractical to mingle Marcellus and Utica fluid. Expensive techniques known as smart wells could be employed, but the more likely option would be to produce them sequentially. Utica would go first and later the vertical portion of the same well bore could possibly be used for the Marcellus.

Another distinguishing feature of the Utica is the presence of a significant oil portion on the western side. Much of this will be in Ohio, a relative newcomer to the shale bonanza and one in dire need of job creation. In fact the field it is most similar to is the Eagle Ford Shale. It has the same pattern of hydrocarbon distribution: oil, wet gas, and dry gas. The Eagle Ford has been the fastest growing shale area by far as measured by the rate of increase of production (see Figure 7 in chapter 12); one could likely expect the same with the Utica. Chesapeake Energy reported huge results from wells in Ohio. They reported gas production to be between 3 and 9 million cubic feet per day (MM cfd) and with associated natural gas liquids (NGL) of 800 to 1,500 barrels per day ("Chesapeake's New Utica Shale Wells," 2011). The ratio of liquids to gas is similar to the numbers in the Marcellus, but the raw numbers are much greater, as would be expected from the greater depths. But those reports are from 2011. More recently the news is more tempered, although many players have now entered. Shell reported recently (Royal Dutch Shell, 2014) that two dry gas wells in the Utica were highly productive, with peak production in one of them in the vicinity of 26 MM cfd, which is huge (Royal Dutch Shell, 2014).

The Oil/Gas Price Spread and the Effect

In the early part of the last decade, oil and gas prices were in substantial parity. A barrel of oil has about six times the energy content as a thousand cubic feet of natural gas. So, if the prices are in parity, the ratio of price of oil in dollars per barrel to natural gas in dollars per thousand cubic feet should be around 6 to 1. The last sustained period for that price ratio was back before 2007. The trend since then has been for the ratio to be in the mid-20s to 1, and 2013 estimates from the EIA (US EIA, 2012, January 23) forecast this ratio to stay in that range for two decades.

This is entirely due to shale gas. The supply has kept the cost down. At the same time, oil has continued to become scarcer and its prices have stayed up. Because of this spread between oil and gas prices, the oil and wet gas are the most desirable. One would expect these to get developed first. This means the western side of the Utica will be advantaged economically, at least at first.

The Marcellus is similarly wet on its western portion. Taken together, one would expect the greatest activity to be in western Pennsylvania, West Virginia, and eastern Ohio, and in 2014 that is in fact the trend. The last two are the current have-nots of the shale gas boom. West Virginia will likely be somewhat ambivalent about this because of the prominence of the coal industry there. But Ohio has welcomed it. The recession has hit that portion of Ohio particularly hard. Of course, other considerations will be in play, including whether the liquids processing is done in state or elsewhere.

The propane and larger molecules will have an instant local market. Half of the liquids comprise ethane (ethane is a gas at normal pressure and temperature but falls in the classification of NGL). The dominant use for ethane is in converting to ethylene, which in turn is used to make a host of products such as polyethylene. No capacity for this process exists near the areas of production mentioned. A possible solution to that problem is discussed in chapter 14.

PART II

Environmental Issues

II. Environmental Issues

To many readers this section will define the "peril" in the title, although pitfalls other than directly environmental are peppered throughout the book. In my view, the most substantive environmental risk rests in the improper surface handling of components of fracturing fluid and subsequent disposal of the fluid returning to the surface (flowback water) following the fracturing operation. Baseline testing prior to any commercial activity is paramount in determining causality.

The issue of full disclosure of chemicals has been highly visible with the public. Virtually every state regulation includes a controversial provision to permit non-disclosure of this information if deemed a trade secret. Freshwater usage is a flash point as well, especially in drought conditions, as each well uses up to 6 million gallons of water.

Earthquakes related to shale gas production activity have also caught the attention of the public. In early 2015 the US Department of Energy issued a report confirming that earthquakes can be generated during deep disposal of waste water. The problem lies in the deep disposal. The mechanisms are understood, and I discuss measures that could be taken to avoid these occurrences.

CHAPTER 5

Beyond Gasland

"We can work it out"

—From "We Can Work It Out" by The Beatles (written by John Lennon and Paul McCartney)

No shale gas production issue may be more fraught with partisan rhetoric than the possibility of methane in drinking water. Flaming faucets make for great imagery no matter the true frequency of occurrence.

Well water contamination is very personal and frightening. Think *Erin Brockovich*. Airborne species don't appear to get the same reaction. Certainly, carbon dioxide in the air barely registers on the average personal anxiety scale. Consequently, assaults on the quality of well water make for avid reading and activism. In the case of shale gas, industry response to well contamination has been sweeping in denial. Both sides play fast and loose with the English language, as will be shown.

There are two potential ways in which shale gas operations could contaminate aquifers. One is through leakage of the chemicals used in fracturing. These would then be liquid contaminants. The second is the infiltration of aquifers by produced methane. This is a gaseous contaminant, albeit in the main dissolved in the water. If present, a portion may be released as a gas, as spectacularly depicted in the documentary *Gasland*. Natural occurrences such as the Eternal Flame Falls in the Shale Creek Preserve in New York demonstrate methane intrusion into a freshwater source. The name follows from the fact that the gas remains lit with a visible flame under the rock overhang of the waterfall.

Incidentally, methane is odorless and colorless, so it is hard to identify. It leaves no taste in the water, but can be hazardous if it collects in an enclosed space. When used in commerce, industry deliberately adds mercaptans, a smelly substance (travelers on the New Jersey Turnpike proximal to refineries in years past know the smell well) for added safety. Gas leaks in a commercially supplied setting such as a kitchen or furnace are therefore detectable simply by smell. Leaks from a well would not be detectable by smell.

Natural contamination is either from relatively shallow biogenic methane or from thermogenic gas from deep deposits escaping up along faults and fissures. These last are generally due to tectonic activity at some time. The two types of methane gas have fairly different fingerprints and can often be distinguished on that basis. Good oil and gas exploitation practitioners will avoid producing in areas with significant vertical leak paths because they vitiate normal sealing mechanisms.

Distinguishing Between Methane Sources

Thermogenic methane is formed when heat and pressure act upon organic matter present in the sediment. The stress of this "thermal maturation" causes the longer molecules to be chopped up to form smaller molecules. The smallest is methane and usually dominates the mix. Also present will be the somewhat larger molecules such as ethane and propane, collectively known as natural gas liquids (NGLs). Thermogenic methane will almost always have some of the larger species present.

By contrast, biogenic methane will almost never have these other gases. This is because the method of formation is bacterial action upon carbon-bearing molecules such as carbon dioxide and hydrogen, yielding methane molecules, water, and energy. This mechanism is unlikely to form bigger molecules.

The absence of larger molecules is generally a clear indicator of biogenic origin. The reverse is not necessarily as clean an argument in part because mixtures of gases from different origins are possible.

Another distinguishing feature is the isotope signature. The scientifically faint of heart can skip this paragraph without loss of the thread. Carbon has a heavy isotope ^{13}C, with an extra neutron compared to the normal species. When methane is formed by either mechanism above, the carbon atoms in the methane preferentially have a higher proportion of light atoms compared to the source material. This is known as fractionation in that the heavy atoms are preferentially left behind. But bacteria tend to want to feed on lighter carbon and so the fractionation is more pronounced than it is for thermogenic reactions. Also, the stuff they feed on is often of bacterial origin so is already concentrated with the light version. So, the fractionation is an indication of origin.

The isotope signature, taken together with the level of presence of larger molecules, is the generally accepted means for distinguishing between gas origins.

Despite the seemingly sound scientific fingerprinting techniques available, identifying the source of gas is subject to interpretation. This is particularly the case when, within the class of thermogenic origin, efforts are made to identify the age of the rock from which the leakage took place. In principle, both the isotope signature and the NGL content have relationships with rock age. These relationships are not always well behaved enough to be unequivocal.

Water Well Contamination Studies

The first comprehensive study (Osborn et al., 2011) of possible contamination of water wells from shale gas production activity was by Osborn and colleagues from Duke University. They studied 60 wells in Pennsylvania and found no fracturing chemicals in the water wells. They did report finding a strong correlation between proximity of shale gas wells and methane in the well water. The data have been interpreted by the authors as proving the origin of the methane to be from the Marcellus formation from which the gas was being produced. The method used was as described in the box at left. Since no baseline testing of the water wells was done, the entire conclusion rests upon the method of analysis. Any retrospective study faces this shortcoming.

More recently, industry proponents took the same data and published an analysis to the effect that

> the isotopic signatures of the Duke study's thermogenic methane samples were more consistent with those of shallower Upper and Middle Devonian deposits overlying the Marcellus Shale. This finding indicates that the methane samples analyzed in the Duke study could have originated entirely from shallower sources above the Marcellus that are not related to hydraulic fracturing activities (Molofsky et al., 2011).

They based this in part upon other studies they conducted in which they tested 1,700 wells. As a clarification of terminology, the geologic periods named above span about 30 million years—not huge in geologic age terms. That is why they were careful to use the conservative phrasing "could have" in relation to their assertion. They, too, did no baseline testing and note that the area is well known for methane intrusions in freshwater aquifers.

They provide a detailed description of the geology of the area and identify many small gas bearing formations. In my recommendations below I suggest regulations requiring the identification and sealing of these gas-prone intervals in the shale gas wells.

Such dueling interpretations are inevitable when dealing with data with no baseline studies. The only rational way to determine the propensity for methane contamination attributable to gas production is to conduct baseline analyses of water wells prior to any activity. Sound inquiry in any field of science has always had this feature. A paper on biology experiments would be rejected in a peer-reviewed journal if these "controls" were not present. Both sides agree with this, and in fact the EPA is conducting such studies as part of the congressionally mandated investigation of fracturing operations. The most thorough type of study would be one that did all of the above and also considered the geology and the integrity of gas well execution. The US Geological Survey (USGS) conducted a study in the Fayetteville formation in Arkansas (Kresse et al., 2012). They used baseline testing and concluded that neither methane nor chemical contamination occurred. Incidentally, three of the authors in the Osborn study (Osborn et al., 2011) co-authored the USGS report.

The distinction between potential liquid and gaseous contamination is important because the hazards are different, as are the remedies and safeguards. Also, because well water could not naturally have the liquid contaminants from fracturing fluid, any presence at all is evidence of a manmade source. Therefore, simple testing of wells proximal to drilling operations is sufficient, with the only possible complication being some source other than drilling, such as agricultural runoff. This is easily resolved because of the specificity in the chemicals used for fracturing. The fact that Osborn et al. (2011) found no such chemicals in any wells is hardly ever reported in the press, indicative again of the polarization caused by the issues.

Unfortunately, liquid and gaseous contaminations get lumped together in the statements by shale gas opponents and the genuinely concerned public. Some see methane intrusion as proof of well leakage as a whole and therefore equate it to chemical contamination as well. *Gasland* reports "thousands of toxic chemicals" as the hazard, which is hyperbole. In actuality, the mechanisms for possible leakage of gas and liquid are quite different. Methane as gas is much more likely to leak out of a badly constructed well than is a liquid.

So, do producing gas wells sometimes leak into freshwater aquifers? The answer is yes. In all cases this is because of some combination of not locating cement in the right places and of a poor cement job. Many wells will have intervals above the producing zone that are charged with gas, such as the formations identified in the *Oil and Gas Journal* paper cited earlier (Molofsky et al., 2011). If these are not sealed off with cement (many wells are not cemented top to bottom by design) some gas will intrude into the well bore. This will still be contained unless the cement up near the freshwater aquifers has poor integrity. In that case the gas will leak. You will notice nothing in the prior discussion says anything about fracturing. In other words, a badly constructed well is just that, no matter how the gas was released from the formation. This distinction is lost on many. The paper by Osborn et al. (2011) unequivocally shows no fracture chemical intrusion into water wells. It also shows gas intrusion in disturbingly many cases, although no baseline measurements were made to normalize for possible natural seeps and prior drilling activity. Yet the title of the paper is "Methane Contamination of Drinking Water Accompanying Gas-Well Drilling ***and Hydraulic Fracturing***" (emphasis added). The last three words infer a causality that is not proven and in fact is contraindicated by the absence of fracturing chemicals in the water wells.

Industry proponents on the other hand make statements such as "hydraulic fracturing has never contaminated groundwater." In precise terms this may be right in that fractures have not propagated into groundwater. Take the hypothetical case of a well associated with fracturing operations that leaks gas but not liquid. One could argue that the poor construction would simply not have occurred but for the desire to fracture the shale reservoir to produce the desired fluid. So an opponent would take those very data and say "hydraulically fractured wells contaminate groundwater," while the proponent could say "hydraulic fracturing did not contaminate groundwater." Neither would be wrong. It is the public that will be confused with this license taken with the language.

Suggested Remedies

Rhetoric aside, we can take proper stewardship of our resources and the environment. Some possible measures are listed here and policy suggestions are made in chapter 30, "Policy Directions." Permits must be given only to oil companies with good track records, thus maximizing the chances of diligence in well construction. Water wells proximal to intended operations (the Pennsylvania governor's Marcellus Shale Advisory Commission recommends 2,500 feet) should be tested prior to drilling, at the cost of the operator. The state Department of Environmental Protection should maintain a list of certified testing laboratories and only these ought to be used. North Carolina rules awaiting ratification by the NC legislature in 2015 have all the above provisions and additionally require periodic testing of the same wells.

Logging ought to be required to identify zones of possible minor gas production. Logging is a routine technique in which specialized sensors are lowered into the well to identify formations of interest. The state Department of Environmental Protection must require that these zones of gas be sealed with cement. Consideration may be given to requiring cementing top to bottom in areas whose geology dictates that. At a minimum, the cement bond log ought to be required. This examines the integrity of the cement job and famously was not run on the *Deepwater Horizon* well that blew up in the Gulf of Mexico.

Routine testing of the water wells ought to be de rigueur, with a prompt attempt to seal the well if leaking. This occurrence should also result in a severe penalty. Regulations requiring all of the above will not be considered punitive by the industry. This is because normal sound practice includes all of these, with the exception of the stipulations with regard to water well testing. That is not done routinely, although there are several instances of this happening voluntarily. All of this and adherence to sound drilling and completion practices are necessary to ensure the sustainable production of a valuable resource.

CHAPTER 6

The Chemicals Disclosure Conundrum

"Because you're mine, I walk the line"

—From "I Walk the Line" by Johnny Cash

This earth is ours. We have an obligation to it and the people who inhabit it. The shale oil and gas industry can do right by it and still conduct commerce. This is especially so in the matters pertaining to using chemicals wisely. Most states in the US require the disclosure of chemicals used in fracturing fluids by posting on the public website FracFocus.org. However, the majority of instances include the so-called trade secret exclusion. In essence this permits the operating company or its subcontractor to withhold information if it deems the chemicals in the fracturing fluid to be a trade secret. This is one of the most contentious issues surrounding shale oil and gas regulation. The public appears concerned that under the veil of trade secrecy, industry could use chemicals harmful to human health. Some of the concern is as simple as the need of the public to know what goes in the ground.

Trade Secret Exclusion

Why should there be trade secret exclusion? Regulators often are asked this question. One reason is that legitimate trade secrets are protectable under many state laws, as for example the North Carolina Trade Secrets Protection Act, embodied in Article 24, Section 66-152 of the North Carolina General Statutes. The other reason concerns the public desire for the industry to create means to use greener chemicals. If a company succeeds, it ought to be afforded intellectual property protection rights, the sort available to all citizens. One means for such rights is patents. The neat feature about patents (from a public disclosure standpoint) is that 18 months after filing, all the contents are published whether or not the patent is eventually granted. Rarely does an invention make it to commercialization within that period. So the public will in effect be informed in good time.

Intellectual Property

Intellectual property in commerce falls into four principal categories: patents, copyrights, trademarks, and trade secrets. The first three are covered by federal law in most countries. Trade secret law in the US is at the state level, although most states follow the same strictures by using the Uniform Trade Secrets Act.

Patents are granted to a person or entity if certain standards of novelty and non-obviousness are met. This entitles the holder to exclusivity for 21 years following the filing. In return for this right, the public is allowed access to the invention after the 21 years. The framers believed that this compact was best for the world at large because a closely held secret inhibits commerce due to the fact that only one entity can supply it and innovate to improve it. A common example of this value to the public is the plummeting cost of a drug when it goes "off patent." Generic drugs constitute a huge and thriving business affording the public low-cost access.

As mentioned in the main text, all patents are published 18 months after filing. This was not always the case in the US. Inventors had been allowed to file and had no requirement to publicly disclose the contents were the patent not granted. But the comfort granted the public by this disclosure is tempered by an interesting aspect of clever patenting. The best patents, from the point of view of the inventor, are those in which enough information is provided to the examiners to get allowance, and yet critical elements are withheld to prevent competitors from knowing what exact combination was used or in what precise proportion. In the context of public disclosure of fracturing chemicals, a point of note is that these stratagems are devised to thwart competitors, not to deny the public information. But they could nevertheless have that effect, and state regulations ought to take these nuances into account. However, as noted later, from a public safety standpoint, not knowing proportions may not matter much.

Trademarks and copyrights are types of property that have no bearing on the main debate in this chapter. But trade secrets certainly do and are, in fact, the main issue. Inventions are often kept as trade secrets if they could easily be worked around. This is often the case for apparatus as opposed to methods. Also, patents are not useful unless the infringer can be detected as doing so. So, inventions that do not meet these two standards will be considered for relegation to the trade secret pile. The point here is that there are good and valid reasons for holding innovations as trade secrets, and doing so does not necessarily make them any less innovative. Regulators ought to treat trade secrets with as much respect as any other form of intellectual property. Accordingly, the trade secret exclusion to the chemical disclosure laws has a legitimate place.

Sometimes a company may choose not to patent and to merely hold the innovation as a trade secret. An example is the formula for Coca-Cola syrup. The formula constitutes intellectual property and it would not be fair to require disclosure, which would enable their competitors to copy them. This appears to be the basis for the trade secret provisions in state laws. In these cases, companies have stringent procedures to prevent inadvertent public disclosure. Any company claiming the trade secret exclusion ought to be required to submit an affidavit asserting that the claimed item received such care and that the details had not already been made public. This requirement could be expected to limit the exclusion claims to genuine trade secrets. The foregoing forms the basis for the rule set by the North Carolina Mining and Energy Commission, of which I am the chairman. It was passed by the North Carolina legislature and is now the law.

Some in industry are pushing back on the idea that trade secret claims face verification by a state body. One reasonable basis for concern on their part would be undue legal burden, with attendant costs and delays. This concern would be taken care of by and large if the state body merely required the filing of a short brief stipulating that the claimed trade secret had received the same care as their other trade secrets. In unusual cases the state body may seek a short interview to clarify some aspect. The second area of concern could be the inadvertent disclosure of the trade secret by the review body. This concern can be addressed in a couple of ways: The review body could have very few people, as few as three. The secret information, and all documents pertaining to it, could be returned to the company promptly after the review. This sort of thing is commonplace in commercial transactions such as the sale of one company to another. The business confidential information in these cases may have much greater value than those in the matters discussed above, and yet the secrecy aspect of the transaction is rarely a consideration impeding the deal. Furthermore, most state agencies routinely receive business-sensitive information in the process of permitting or the granting of special considerations such as tax forgiveness.

So, What Are These Trade Secrets Anyway?

Without exception, the trade secret exclusions are sought by the service company or a supplier to them, not the oil and gas company that will use the products in the well completion process. In many cases the oil company is in the dark regarding the precise formulation. But increasingly the medium to larger oil companies are asserting their purchase power rights to demand fuller disclosure. For example, according to a 2014 declaration on its website, Apache Corporation now requires the "elimination of diesel, BTX, endocrine disruptors, and carcinogens" as constituents in fracturing fluid. BTX stands for benzene, toluene, and xylene, all of which are known as aromatic organic molecules, originally so-named because they had an aroma. *Aroma* in common parlance tends to be associated with pleasant smells such as from perfume. The "aroma" from BTX compounds is more like a bad smell; take a cautious sniff sometime when you are filling gasoline and you get the idea.

Volatile Organic Compounds, or VOCs

Volatile organic compounds are broadly classified. Basically, any organic compound that has noticeable vapor pressure at ambient temperature falls into this category. Some are benign, such as those generated by plants, and some are toxic. The rest fall somewhere in between. The organic solvent in paint is one such. In fact, many household materials, including cleaning liquids, emit VOCs. Some of the compounds are toxic, and their impact on human health depends upon the level of exposure. Common compounds found in the home are benzene, toluene, xylene, formaldehyde, and acetone. All are harmful at some level and some are carcinogens.

The debate on VOC levels associated with shale gas centers firmly on the toxic and potentially carcinogenic species. Much of the public angst from the standpoint of disclosure comes from a concern that such compounds could be used in secrecy, shielded by trade secret exclusions.

Oil company declarations regarding what they want and don't want as fracturing fluid ingredients are important because service companies know that the customer has a choice. This is especially so in shale gas operations, which use "slickwater" formulations containing fewer chemicals than in more conventional fracturing operations. Everybody pretty much uses the same chemicals. They may use somewhat different formulations to assert differentiation.

So, the *recipe* may be the secret. If that is the case, it is not a public health issue because the *ingredients* themselves are fully disclosed. Why is the nondisclosure of the recipe not a public health problem? It is not an issue because the *proportions* of what went into the ground are not terribly relevant. What goes in is not what comes out. Some constituents are consumed, such as biocides, some are partly reacted, such as acids, and some return in the form introduced. So, more important is to analyze the water that flows back and not worry too much about the *proportions* of what goes in. The *ingredients* that are going in inform us of the reaction products that may emerge, and we can analyze for those in the return fluid (flowback water).

The Way Forward

The use of standard chemicals allows for high-quality shale gas wells. Companies are becoming increasingly open on this point because of public concern with nondisclosure. Halliburton has posted the precise ingredients it uses on its website (Halliburton, 2015). I examined the Marcellus list, and each chemical has a Chemical Abstracts Service (CAS) number. This number uniquely identifies the chemical, and the properties are easily discerned. Virtually all service companies use some variant of these very chemicals. There is no pressing need for substitution of these with others except to make them greener—more on that below. The trade secret exclusion would apply only to substitutes with use advantages. Such a claim would be very hard to substantiate in shale gas wells; the standard chemicals work just fine. However, I recognize that this fast-moving sector will be in continuous improvement, especially in recovering a higher fraction of the hydrocarbon in place. This is particularly the case in liquids-rich plays and also shale *oil* wells in the Bakken and elsewhere where only about 5 percent of the oil in place is currently being recovered. But the improvements must not compromise the environment or public health. Certainly compounds such as those in the BTX family ought to be disallowed by law. The North Carolina rule set forbids any aromatic compound as a constituent. This is a good rule and does not constitute an imposition on industry because relatively benign alternatives are available. Full disclosure of the CAS numbers for substitutes ought to be the goal. Oil and gas companies have a lot of choice when it comes to service providers and ought to be discerning on this point.

The goal is for functionally equivalent chemicals with minimal environmental impact. Norway has been using this approach for at least a decade—requiring green, yellow, and red labeling for oil field products, and requiring a game plan (Norwegian Petroleum Directorate, 2010) to reach full green slates, meaning all constituents in the operations are green. One oil service provider has come up with a methodology for this for fracturing fluids. Another such provider has a fracturing fluid composition fully comprising additives from the food industry. Their CEO famously drank some quantity of the fracturing fluid (we assume the sand was absent) on a public occasion. He has since been seen in public. Incidentally, if my recommendations in chapter 7 of using saline water are followed, he should take a smaller sip. Then again, who among us has not taken that inadvertent mouthful of sea water while body surfing? The astute majority who don't body surf, that's who.

Clearly this sort of innovation ought to be granted trade secret status, if sought, so long as the requirements of the relevant state statutes are met. But since by their very nature these ingredients will be environmentally benign, the secrecy will almost certainly be in the production processes to make them cheaper or in the recipe. As discussed above, the recipe ought not to be of concern to the public, especially when benign ingredients are involved. To the extent the secret is in the green chemical ingredient, it ought to receive trade secret status after some verification.

In conclusion, oil and gas companies ought to strive to employ only those service companies prepared to fully disclose the chemicals used, including the CAS numbers of each. In every case the state ought to accord the recipe trade secret status if claimed, without subjecting it to any process for verification of claim.

CHAPTER 7

Preventing Contamination of Surface and Groundwater

"A bridge over troubled water"

—From "Bridge Over Troubled Water" by Simon & Garfunkel (written by Paul Simon)

The safe handling of water associated with shale gas operations is the most significant environmental hurdle faced by the industry. This can and must be surmounted. This is not to downplay fugitive methane emissions or any other concern, but this one is up close and personal to the community surrounding gas production. The global warming propensity of methane, and arguments regarding the different models to assess the impact—these simply do not have the resonance of potential short- and medium-term health effects of polluting surface or groundwater. These last are what one could call front-of-the-box items.

The potential for pollution of surface water and groundwater comes from two sources. One is the improper handling of the return water from fracturing operations, known as flowback water. The other is from accidental spills of all sorts, including fracturing fluids and water laden with these chemicals. The very nature of shale gas formations is such that even if fresh water is injected as the fracturing fluid, the returning water is much more salty. The salt content could be as much as 350,000 parts per million (ppm). Sea water is about a tenth as salty, and fresh water is defined as less than 500 ppm. The saltiness is due to the fact that the returning water includes water resident in the formation, which happens to be heavy brine.

Radioactive Elements

Naturally occurring radioactive elements can be found in the flowback water. These can be isotopes of uranium, thorium, potassium, or radium. *All* shale strata have some amounts of radioactive species naturally. In fact, in the prospecting for hydrocarbons, shale is distinguished from sand and

The Front of the Box

A 2011 *New York Times* story (Kaufman, 2011) has a very interesting take on the environmental movement and changes therein. These organizations in the past have taken national or even global approaches to the issues. The rise of global ambient temperatures caused by greenhouse gases is a case in point.

The general public can be left cold at two levels. One is that global issues do not resonate with a lot of folks, while local ones do. The other is the discounting of future privation. This is not unlike discounting future earnings in finance; a discount rate is applied which gives a lower present value. Similarly, future suffering is discounted, especially when it is 40 years out, as are most global warming warnings. Rising water levels on a Florida beach 40 years hence (and only a maybe at that) have little resonance with the public in Wyoming. One could call it two degrees of separation.

The *Times* story draws a clever analogy. If a consumer is walking down a grocery store aisle and sees a box with a delectable brownie on the face, she may be attracted to it. Some might look at the back of the box, which details the information indicative of a potentially obese future for the consumer of the goods. Even though the future in this case is more in the short term than the aforementioned global warming one, the choice of looking at the back of the box—at the potential hazards—is personal and will not happen all the time.

The *Times* story concludes that environmental activism is best served by presenting front-of-the-box issues at a local level, and leave the back-of-the-box issues to the national or global level.

carbonate by the presence or absence of these elements, using Geiger counters in some cases. As discussed in an earlier chapter, hydrocarbons are most usually found in these other two mineral strata, not shale, and so identifying them is a key. Yet these concentrations are usually too small to be a health concern, and usually are below EPA threshold levels (King, 2012). The greater incidences, particularly of radium, appear to be in the Marcellus Shale of New York. Although usually low in concentration, if a crusty layer known as scale is formed anywhere in the system, these elements will have a tendency to concentrate in the scale. Consequently, scale formation must be inhibited. This is why in the chapter encouraging the use of saline water in place of fresh water I suggest the removal of the ions of calcium, magnesium, and barium because these form adherent scale, especially barium.

If scale formation is inhibited, radioactive elements are not likely to be an issue (King, 2012). This is particularly the case if one reuses the flowback water, as I propose later in this chapter as a preferred option. Then any of these elements that come up go right back down where they came from.

Fracturing Chemicals

Rock is fractured by injecting a fluid, usually water, at very high pressures. This operation is made more efficient if the water contains a thickening agent to make it more viscous. This is usually a sugar derived from the guar gum seed, a crop that is largely sourced from India and Pakistan. To further improve the viscosity the molecules are bound together with a chemical known as a *crosslinker*.

The viscous liquid under high pressure creates fractures in the rock. Then, in order to remove the liquid, another chemical is introduced known as a *breaker*. It breaks the cross-linking bonds. Now the liquid is no longer viscous and can be pulled out. A final step before the viscosity is dropped is to inject something known as *proppant* into the cracks. Proppant is usually sand particles but can also be synthetic ceramic materials. Its purpose is to hold the cracks open after the high-pressure fluid is removed. Proppant functions in a manner similar to the pillars and beams in coal mines, which enable the transport of coal-laden bins. As in the case of the mines, if the rock is not propped up in some way, the weight of the sediment above the zone will close the cracks.

Finally, after the fluid is removed, the gas will flow through the propped-open pathways into the main well bore and then up to the surface. Some gas will start flowing, mixed with the fracturing fluid as it is removed and the pressure drops. This mixture will need to be separated at the surface. This early gas, if not piped somewhere, is the primary source of fugitive methane into the atmosphere, as discussed in chapter 9.

Slickwater Operations

The vast majorities of shale gas operations do not use the thickening agents and associated chemicals. This is because the early production in the Barnett Shale, in Texas, had dramatic falloff in production over time that was attributed to the plugging of the cracks by residues from the sugars used for thickening.

This use of essentially fresh water is known as *slickwater production*. Despite the imagery of slipperiness from the name, the fact is that plain water is not very slick at all and the friction in the pipes can be high. So, a friction reducer is added, which is usually a polyacrylamide, the same material that is used in baby diapers, wound dressings, and the like as an absorbing material. Also, simply greater volumes are needed in this case. On average about 4 million gallons of fresh water are used per well, but the use of 6 million gallons is not unusual.

Chemicals in Slickwater Fracturing Fluid

- **Friction reducer:** Always used. Usually polyacrylamide. Other common uses: baby diapers, flocculent to remove fine particles in drinking water preparation.
- **Scale inhibitor:** Used about a quarter of the time depending upon solutes present. Usually phosphonate. Other common use: detergents.
- **Biocide:** Used in almost every instance. Glutaraldehyde, chlorine dioxide, mostly the former. Other uses: glutaraldehyde as medical disinfectant and chlorine in municipal water supplies. Will be replaced in part with surface treatment designed to kill the bacteria using ultraviolet radiation.
- **Surfactant:** Sometimes used. Many formulations. Other common uses: soaps, cleaners.

Other minor constituents of the fluid include scale inhibitors, for the reasons mentioned above, and biocides. This last is to address the fact that bacteria of certain types are harmful to the operation. Some species cause the formation of sulfur-bearing gases. In fact virtually all hydrogen sulfide found in reservoirs was formed by bacterial action. Also, certain other bacteria cause the formation of salts, which plug the pores and impede production. For all these reasons operators will not permit any bacteria in the fluid injected and additionally inject biocides, the principal one in use being glutaraldehyde. Other methods in some use include the use of ultraviolet radiation to kill the bacteria in the fluid prior to injection.

Use of Diesel and BTEX

One of the most vigorous arguments against shale gas production has been about the use of diesel in fracturing fluid. *There is no technical reason to use diesel in the fracturing fluid in shale oil or gas operations.* When diesel was used in the past it was as a lubricant. Now polymers do the job well and are safer to handle. *The use of diesel for this purpose should not be permitted*, and such a rule will not impose any material hardship on operators. The class of organic compounds known by the acronym BTEX has also been cited as in use.

These compounds may well be present in diesel, but if that is outlawed, BTEX constituents from that source should cease to be an issue and BTEX ought to go away as well, given there is no technical need for them to be used in isolation. The Nother Carolina regulations forbid any aromatic compounds to be used in fracturing fluid. BTEX compounds all fall into the aromatic classification.

Some reservoirs are water sensitive. In one type of sensitivity, the formation swells with water and impedes the flow of hydrocarbons. In this case, which is rare, propane or liquefied petroleum gas (LPG; a mix of propane and butane) may be used. While diesel use is a possibility as the carrier fluid, there is no need for it since these other two work better at about the same cost. At a lecture I was asked by an environmental group official why industry did not use propane in place of water. My response was that to use an expensive hydrocarbon to produce another was something most would prefer to avoid. Keep in mind that propane pricing is pegged to oil, which is about four times the price of natural gas. Transporting and handling propane for this use would open new concerns. There are some technical advantages to using liquid propane but they do not balance the cost and risk. An added element of cost is that the propane will blend with the produced natural gas and will need to be separated out. This requires cryogenic apparatus at the rig site. Because of these and other issues with alternates, water can be used safely as a fracturing fluid, and ought to be.

Diesel is also predominantly used as the fuel for the pressure pumps on the rig. The issues with this would involve leakage, spills, and emissions. Pennsylvania has expressed concern because much of this activity is in populated areas. The emissions of concern in diesel usage, whether on a rig or on a city bus, are particulates. Metropolitan areas that switched public transport from diesel to compressed natural gas (CNG) have seen dramatically positive health effects. These are reported in chapter 17. In addition to the use of CNG, or liquefied natural gas (LNG) for that matter, another possibility is to substitute diesel in part or whole with dimethyl ether (DME). Diesel engines today can substitute DME to 20 percent or more with no modifications. Engines can run on pure DME if modified. Isolated instances of substitution of diesel are occurring. The leader on this has been Apache Corporation, with the service assistance of Halliburton and the equipment support of Caterpillar. The ultimate goal would be to use field gas directly or with some processing. States should take a good look at DME and CNG (or LNG) substitution and take steps to encourage and enable these changes. In the grand scheme of environmental hazard, these are small potatoes. But you want to take the wins where you can get them. This one really has no technical hurdle, just the will to do it.

Disclosure of Chemicals Used

In one sentence, *full disclosure must be made of all chemicals in fracturing fluid.* The only exception ought to be the use of a proprietary product, and for that exception to be given *the bar must be set very high.* This entire issue is covered in greater detail in chapter 6, "The Chemicals Disclosure Conundrum." But here are just a few points for context with this chapter.

Even in such a case the properties of the material relative to environmental issues must be fully documented. There are legal methods for confidential disclosure to federal and state authorities that protect all parties. If a patent is applied for, 18 months following the application the US Patent and Trademark Office will publish all the secret details anyway. The oil field is famously slow to commercialize new products. An 18-month stay of disclosure is virtually in the noise. The proprietary product exception must not be a shield or an artifice.

The mandatory disclosure that is noted above is for the gross chemical, say polyacrylamide. No chemical is 100 percent pure. To identify each possible impurity would be onerous and may in fact not be easily performed. As long as the impurities have no material bearing on public health, their disclosure ought not to be mandatory.

In the run-of-the-mill shale gas operation, the chemicals used are well known, as described in the box on page 44. Note, however, that in some wet gas operations, and almost all shale oil operations, additional chemicals are added, as described above. Primarily, these are a thickener, such as a derivative of the guar seed, a crosslinker, and a breaker to remove the crosslinking action. Non-disclosure by companies is likely on the advice of attorneys who fear the famous "slippery slope" concept. The way that goes is roughly analogous to that other well-known adage "give an inch and they will take a mile." Advisors feel if you disclose something now, even more will be asked for later and so on. Memo to attorneys who feel that way (and by the way not all do; this is not a diatribe on attorney conduct. Some of my best friends . . .): without risk there is no profit. Second memo: when non-disclosure of something benign causes folks to get riled up, that is bad business.

One may well ask why allow such an exception at all. The answer is that we want innovation to design the most benign additives at the lowest cost. The reason diesel was used as a lubricant early on was that designers were not attuned to environmental concerns. The same goes for why fresh water is used as the base fluid rather than low-value saline water. We need to challenge the entire industry, and this goes beyond the oil field, to design with an eye to sustainability: environmentally benign and using the least energy, while still being profitable.

If the industry is really clever about all this, it needs to be afforded the protection offered by law in the patenting process and to some degree in the trade secret maintenance process. Premature public disclosure kills patent validity. Patents will simply not be granted, especially abroad. We want the innovation and the associated profit to happen. In context we want it in the area of green chemicals and processes (Heintz & Pollin, 2011).

Handling Flowback Water

Some of the water and chemicals injected into the reservoir return as flowback water. Less than 10 percent of the chemicals injected return to the surface (King, 2012). The polymers degrade at reservoir temperatures, the biocides do their job and get consumed by bacteria, and other chemicals are partly trapped in the rock. Much of the water also stays behind. Recent theories suggest that water actually acts as a proppant in some fractures. In any case only between 16 percent and 35 percent of the water comes back to the surface. Although very little of the injected chemicals return, the flowback water now also carries with it some of the mineral content from the reservoir. The chief of these are chlorides of sodium, magnesium, and calcium, which account for the dramatic increase in salinity over what was injected. There may even be some aromatic compounds that are naturally found in association with some hydrocarbon deposits.

Flowback water is absolutely not suitable for surface discharge. It is also not suitable for being sent to municipal water treatment facilities, as is believed to have happened in Pennsylvania. These facilities are not equipped to handle the high salinity or the chemical content. Some have suggested the construction of special facilities to treat the water for discharge. However, the cost of doing that could be higher than that of two measures proposed below.

Deep Disposal

The least costly method of disposal of flowback water or any other wastewater is the deep injection well. This is an EPA-approved method known as a UIC (Underground Injection Control) Class II injection well. The key criteria are the porosity (it should be high) and permeability of the host rock, the need for competent cap rock above it, and firm guidelines on casing and cementing of the well to protect freshwater aquifers. In most states these are monitored by the state environmental protection agency. The host rock is either a porous rock with no fluids or sometimes a rock depleted of fluid by prior production.

This type of disposal is very inexpensive in comparison to the value of the produced fluids. The fully loaded cost including amortization of the well is between $0.25 and $1.00 per barrel of wastewater disposed, with the former figure applying when the well is proximal and owned by the operator, and the latter including hauling some distance, and profit for the injection company. The water hauling is subject to spills since this is done primarily by trucks. Proceedings from a workshop held by the EPA in 2011 (US Environmental Protection Agency, May 2011) has detail on this and other aspects of disposal.

Since the last part of 2011 there was a lot of press about earthquakes in Oklahoma and Ohio presumed to be related to deep disposal wells. I cover those issues in greater detail in chapter 10. But there is no doubt that disposal wells should be planned and executed with care. Seismic studies must establish the absence of significant faults in proximity. There is evidence that some care is already being exercised; the suspect wells in Ohio were shut down promptly. Also, the EPA reports that there currently are more than 145,000 disposal wells in the US today, going back decades. So the occasional problem could have been expected, especially because the earthquake propensity was not recognized, and avoidance was not part of the design. Still, given that it is a regulated and monitored event, any such problem is correctible primarily through requiring that disposal wells not be in close proximity to active faults over a certain size. The specifications guiding this still need to be worked out, and the EPA then ought to add these to their current specifications.

Reuse of Flowback Water

This is the most elegant solution. But in order not to be cost prohibitive, the industry had to tolerate greater salinity. This is discussed in chapter 8. Were we to require fresh water in the fracturing fluid formulation, the desalination alone would send the costs higher. Flowback water salinities in the Marcellus range from 16,000 ppm to over 250,000 ppm. Halliburton disclosed at a CERA conference in 2013 that salinities in excess of 250,000 ppm could be tolerated in specially formulated fracturing fluids. This means no desalination would be needed, except possibly for the removal of the minor constituent divalent ions, for reasons mentioned in chapter 8. Even solids removal may not be necessary. Surface discharge would require a good deal more treatment. So reuse in the fracturing operation would be cheaper than treating for surface discharge. Also, any minor impurities, such as radioactive species, which could be a problem for surface discharge, would present no hazard to the fracturing application. Bacteria would have to be removed because most operators are reluctant to inject these. But this too is being researched and ultraviolet radiation for killing bacteria is already in practice (Rodvelt, 2011). Alternative techniques such as the use of molecular iodine as the kill agent are also feasible

Desalination Methods

The workhorse desalination method is reverse osmosis (RO). This uses a filter known as a semipermeable membrane: it allows only water molecules through and rejects others such as salt. Ordinarily the process of osmosis causes water to flow from the fresh to the salty side, essentially to equilibrate the concentrations. In reverse osmosis hydraulic pressure is applied on the salty side to force water to flow to the fresh side. Depending on the pressure applied, this will cease at some concentration of the brine, usually at about 75,000 ppm salt. At the end one is left with fresh water on one side and very heavy brine on the other. This is commercially used to produce fresh water from sea water in the Middle East, Australia, and other arid places. But the process leaves a brine to dispose of.

In coastal areas desalination operations put the resulting brine back in the ocean, although this practice is in review in some instances due to risk of damage to coral and other species. Certainly, in inland applications disposal will be a concern. Furthermore, this technique is fairly useless when the starting liquid is a heavy brine, as is often the case in shale gas operations.

Forward osmosis (FO) is a relatively new technique and has greater potential because it ought to use less energy. In this case a "draw" solution is designed with constituents that cause water molecules to be drawn to it from the flowback water side. Eventually the constituents causing this action are removed, sometimes by volatilizing them, leaving behind clean water. The flowback water side concentrates into heavy brine, which needs disposal.

A relatively new technique is membrane distillation. This involves a membrane that transmits vapor only. Water vapor can be moved across the membrane, leaving behind the salts. In this case volatile organics, if any, would likely have to be removed first, else they too would go across. The allure of this technique is that low-grade heat can be used for the vapor production. One innovative method combines this with FO.

A popular method is to simply evaporate the water from the flowback water and condense it for use. This is energy-intensive because the latent heat of evaporation has to be provided. The resultant product is very pure. Some outfits are claiming techniques that minimize energy use. Something like this may be needed for the very heavy brines that are too salty for direct reuse and too salty for RO. But evaporation does get more energy-intensive as the salt content goes up.

(Chelme-Ayala et al., 2011). Iodine is at least 20 times as effective as chlorine and is more benign, being classified by the EPA as generally recognized as safe (GRAS).

In the Marcellus and Utica areas, very few strata have been qualified for deep injection. Consequently, this option is largely unavailable. Not surprisingly, this is the area, particularly in Pennsylvania, where the greatest incidents of irresponsible discharge or disposition have been reported. In recognition of this many operators are reusing the flowback water, proving that this is feasible. In many instances the cost will be very low, likely below the upper end of deep disposal cost. In any case, environmentally secure disposal is not optional.

If the costs are prohibitively high, this country will innovate to get the cost down. Or only the more profitable reservoirs will be accessed until such a time as natural gas costs rise to profitable levels for dry gas production. As discussed in chapter 11, wet gas is extremely profitable and will easily sustain these costs of proper water handling. And there is plenty of that in the Marcellus and Utica, the areas challenged by few deep disposal well options, and by far the largest and most prolific plays in North America.

Responsible disposition of flowback water should be mandatory. Until better alternatives are found and fully tested, *the only disposal methods must be UIC Class II deep injection or direct reuse.*

CHAPTER 8

Zero Fresh Water Usage

"Drove my Chevy to the levee, but the levee was dry"

—From "American Pie" by Don McLean

Use of fresh water as the base fluid for fracturing operations is no longer necessary. Salty water with little or no value for human consumption or agriculture can be used. *Salt* water of convenience ought to be the default option in every case.

Shale gas wells use up to 6 million gallons of fresh water per well in fracturing operations. Only up to about a third of the injected water returns to the surface, so more fresh water needs to be added to make up the needed volume, even if the flowback water is reused.

A *Wall Street Journal* article (Gold & Campoy, 2011) tells a tale of water deprivation in south Texas. The town of Carrizo Springs, the imagery of plentiful water notwithstanding, is dealing with the conflicting demands of agriculture and gas development. Oil and gas leases often do not have water rights attached, although in Texas they do. Farmers without oil and gas leases have the option to sell their water to whomever they please. The value of the water is much greater to the oil companies, so they generally get the water they need. In the *Journal* report, data are presented for the number of water wells drilled for this purpose as opposed to all other uses. In some areas of Texas, 80 percent of new water wells drilled are in support of oil and gas operations. In total the number of wells in 2010 is about five times those drilled in 2005. The rate of this activity can only be expected to increase.

Heavy water withdrawals can strain aquifers, especially in drought years. The major gas production areas of Barnett, Woodford, and Eagle Ford in Texas and Haynesville in Louisiana were in exceptional to severe drought in 2011 and continued to flirt with extreme privation through 2013. The prolific Marcellus in the Northeast is spared this particular problem. Nevertheless, water usage has been an issue in some communities, probably because water-related issues tend to have an emotional component. But in the other areas mentioned the issue is quite real. Curiously, the Texas Water Development Board estimates that mining and oil and gas account for less than 2 percent of all water used. The majority is used for crops (56 percent) and municipalities (27 percent). But again, as the new game in town, shale gas drilling will inevitably be a focus of attention because the other uses already exist and are essentially unassailable.

The above numbers notwithstanding, the report of new water wells being dominantly devoted to oil and gas certainly indicates a growing trend. Fresh water drawdown can lead to the need to go ever deeper. At some point the water will get brackish, having salinities in excess of drinking water standards. This is because all groundwater tends to get saltier as wells get deeper. This is why it is rare to find freshwater wells deeper than 500 feet.

While severe droughts are not common in Texas, in the early part of this decade they have been experiencing the worst drought in decades. In general a vertical swath through the country going north from west Texas through Arizona and Colorado and up to Wyoming is a perpetually water-deprived area. Important gas production areas other than those for shale gas are in Colorado and Wyoming. These are also in tight rock, albeit most commonly sandstone. Fracturing is required to economically release the fluids. This co-location of tight reservoirs with drought-prone environments is nature's little joke, as it were.

Salt Tolerance of Fracturing Fluids

The industry has been pursuing the objective of salt tolerance for a number of years. The initial driver was to improve the feasibility of reusing the water returning from the well after fracturing the rock. On the entire Eastern Seaboard of the US, deep disposal wells are not feasible due to the geology. Pennsylvania operations are rife with alleged instances of improper disposal leading to groundwater contamination. Absent a solution, production could well be halted. One solution is direct reuse at the site.

If fresh water was required, the desalination cost would be prohibitive in many instances. This is because the returning water could have salt content in excess of 200,000 ppm. The conventional desalination workhorse, reverse osmosis, is essentially useless because it needs to treat salinities in the vicinity of sea water (roughly 35,000 ppm) or lower, and reverse osmosis *rejects* water at 80,000 ppm. Evaporative recovery, no matter how clever the design, faces the hurdle of supplying energy to overcome the latent heat of evaporation. So, the objective is to maximize tolerance to the salinity of fracturing fluids.

How Fresh Does Water Have to Be?

Freshness is generally defined by the water's saltiness. The measure commonly used is total dissolved solids, or TDS. In the main these are chlorides of sodium, potassium, magnesium, and calcium. Drinking water is required to be under 500 parts per million (ppm). As a frame of reference, 10,000 ppm is 1 percent and sea water runs around 35,000 ppm, although that can vary from sea to sea.

Agricultural uses generally dictate salinity under 1,000 ppm. Curiously, though, the tolerance for chlorides is variable between plant species. At one extreme are date palms, which can handle up to 20,000 ppm, probably because of adaptive mutation to an environment wherein evaporation tends to render much surface water brackish. Sweet sorghum, a potentially important source of biofuel, is said to tolerate 3,000 ppm. Livestock, too, have variable tolerances. Sheep are the most tolerant, coming in at about 6,000 ppm for healthy growth. They can tolerate double that for a maintenance situation.

Salinity tolerances of livestock and poultry
In parts per million of total soluble salts

Animal	Maximum tolerance for healthy growth
Sheep	6,000
Beef cattle	4,000
Dairy cattle	3,000
Horses	4,000
Poultry	2,000

All of the foregoing underlines the fact that potable water is not needed for every application. In fact, the salinity should simply be fit for the intended purpose. This could be important in selecting the most suitable desalination technology.

Also, some saline aquifers could be moderately useful. In fact, one could seriously consider selecting farm products to suit the available water, rather than the conventional approach of treating water to be fresh. Edible plants can be genetically engineered to be more salt tolerant. One extreme is the class of plants known as halophytes, which actually preferentially consume salty water.

Similarly, commercial processes could be modified to accept higher TDS. One such is the fluid used for fracturing operations. More on that on page 46.

Fracturing fluids were originally designed to work with fresh water because such water was plentiful and it was easier to do that. Today salinities of over 250,000 ppm are tolerable for fracturing fluid formulations, with some adjustments to the other chemicals used. Some of the chloride ions impair the effectiveness of the crosslinkers and friction reducers. Of course crosslinkers are used only when sugars are used as thickeners, and this is largely not the case in shale gas production. Nevertheless, alternative chemicals can be employed in the presence of salinity, whenever crosslinkers are required. As mentioned in chapter 7, in the majority of instances the fracturing fluid is water with very few chemicals and no sugars. This is known as "slickwater" fracturing. The slipperiness evoked by the name notwithstanding, such fluids have high frictional losses. So, friction reducers are added. Some of these chemicals are less effective at higher salinities. However, replacement chemicals have been found.

Chlorides of sodium and potassium are particularly tractable. Those of magnesium, calcium, and barium are less desirable. Mostly this is because they will form an adherent scale in the flow system. Scale tends to concentrate radioactive elements if they are present. In some shale gas drilling areas, radioactive radium, thorium, and potassium are found in the formation and may be present in very low concentrations in the returning fluid. While these quantities are generally benign, if concentrated in scale they can present health hazards, especially during scale cleanup.

One solution is to simply remove these elements, known as divalent ions, from the fluid prior to use. The process for accomplishing this is very straightforward and is commonly known as water softening. Most municipal water systems and some homes with their own wells employ this process when the water is known to be "hard." In a domestic situation this is done largely because the divalent ions interfere with detergent action, so soap does not lather up effectively. At any rate, this is a known process and the only consideration is the cost of doing it.

Sources of Non-Fresh Water

To this point I've discussed the use of saline water in place of fresh. In point of fact a number of other sources of water are feasible for use. These include wastewater from industrial processes such as mining, cooling towers, and effluents in general. Samples of these would have to be evaluated and treated for use. A much more practical solution is the use of water from saline aquifers. As discussed earlier, all groundwater gets salty at increasing depths. So, one could expect saline bodies of water to be fairly ubiquitous. This is in fact the case, as exemplified by the map shown in Figure 6, which was plotted by the US Geological Survey about 60 years ago. The white spaces simply represent areas not investigated by them, and not necessarily lacking in such deposits. This is an important point because a lot of the white space is in areas covered by the Marcellus and Utica, both very prolific shale gas regions.

Figure 6. Depth to saline groundwater in the United States

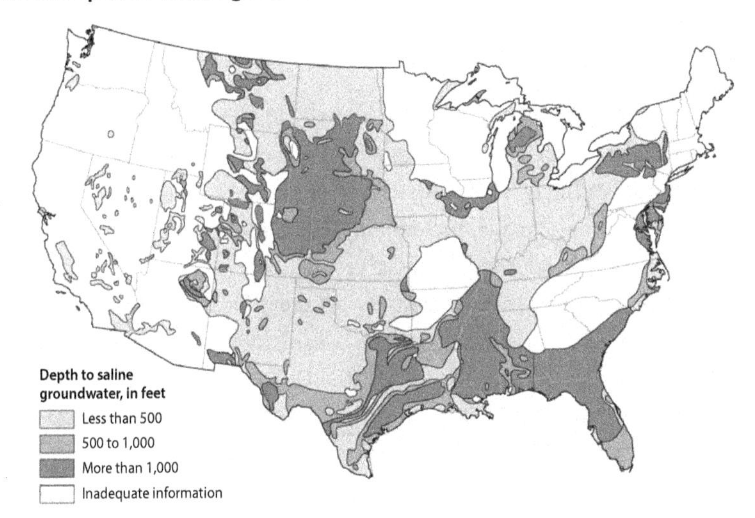

Source: US Geological Survey, 2003, generalized from Feth and others, 1965; Department of the Interior/USGS

Note that in many instances the water is shallower than 1,000 feet, making it very accessible. In general, shallower deposits are less salty. However, in these instances care has to be taken to understand the hydrology relative to adjacent freshwater deposits. In many cases, these are in communication, and withdrawals from one can affect the other.

Saline aquifers can be expected to be in reasonable proximity to producing areas. Consequently they represent reliable sources of water. The other alternatives, such as wastewater, mentioned earlier, while chemically acceptable, could vary in supply. Also, a given saline aquifer can be expected to deliver uniform quality over a long period of time, thus allowing the treatment processes to be standardized.

Assuming widespread acceptance of these ideas, the work needing to be done includes characterization of saline aquifers in producing zones and filling in the white spaces on the map. The principal points of interest are concentrations of divalent ions and bacteria.

Bacteria will need to be eradicated before the water can be used as fracturing fluid. Bacterial species present in the shallow aquifers are likely to be different from those in the gas reservoir. Injection of unfamiliar species could have unintended consequences. We do know that certain bacteria are harmful to the reservoir from the standpoint of causing the formation of chemicals that plug the pores and impede fluid movement, and hence production. Other bacteria can cause the synthesis of hydrogen sulfide. Removal of bacterial species is straightforward. Reuse of flowback water will need some such process anyway. In general, one would expect those process steps to work hand in hand with the preparation of saline water for use.

Horn River Experience

Lest the foregoing appear to be a theoretical exercise, at least one outfit is using these methods routinely. This is Apache Corporation in their Horn River field in British Columbia, Canada. Incidentally, Horn River and Montney in British Columbia are on par with five of the largest fields in the US.

The driver for Apache was at least in part the difficulty of access to fresh water in winter. The water has to travel a considerable distance, and keeping it from freezing is a challenge. They access the Debolt saline aquifer, with salinity running in the vicinity of 35,000 ppm. The bacterial content is low and the mineralogy is acceptable. After minimal treatment, they are able to use it as the fracturing fluid. They sometimes blend it with produced water or even fresh water. They report that water from the saline aquifer provides a cost savings of over 50 percent over the use of fresh water (King, 2011).

The Apache experience is a concrete example of the use of brines of convenience. It would be reasonable to expect that a majority of shale gas operations could virtually eliminate the use of fresh water. From the standpoint of energy policy, this should be the new norm. Departures from this practice ought to require defensible arguments.

CHAPTER 9

Is Natural Gas Indeed Worse for the Environment Than Coal?

"You may be wrong but you may be right"

—From "You May Be Right" by Billy Joel

Until recently, natural gas was seen as an indisputably cleaner alternative to coal. Robert Howarth and colleagues at Cornell changed all that, at first abortively when their study was demonstrably flawed. Their revised report, which now includes the contribution of fugitive methane in coal mining, was published in the peer-reviewed journal *Climatic Change* (Howarth et al., 2011). Their thesis is that about 2 percent of the ultimate recovery of natural gas is released to the atmosphere. Since methane is many times more potent than carbon dioxide as a greenhouse gas, they compute that the net effect is worse than from the use of coal.

Hailstorms of criticism notwithstanding, some of the issues beg debate. A more recent study ("Switching from Coal," 2011) appears to be in support of Howarth and colleagues' contention as well. In contrast is the report by the Worldwatch Institute ("Despite Methane Emissions," 2012), conducted in collaboration with Deutsche Bank, which unequivocally concludes the superiority of natural gas but nevertheless recommends attention to fugitive emissions. Howarth's own colleagues at Cornell University conducted a peer review and pronounced the 2011 publication flawed (Cathles et al., 2012). At last count only one supporting study, cited above, is counterbalanced by at least five in serious opposition.

A blog by Michael Levi (Levi, 2011) of the Council on Foreign Relations is worth reading mostly because Howarth chose to respond to his critiques, and the back and forth is instructive. The most critical report is one by Mary Barcella and colleagues from CERA, entitled "Mismeasuring Methane: Estimating Greenhouse Gas Emissions from Upstream Natural Gas Development" (Barcella et al., 2011). This report alleges that key IHS data were "misused and severely distorted"—strong words. On balance all of the criticism centers on overcounting fugitive methane emissions from shale gas production operations.

So what is the public to make of all of this? They are right to assume that science is deterministic, at least in the broad swaths of the argument in question. When combusted, natural gas produces about 45 percent less carbon dioxide than does coal in producing the same amount of electricity. The Howarth study does not take the analysis to post combustion, thus intrinsically favoring coal because coal has lower combustion efficiency. The salience of such a comparison lies in the fact that cheap shale gas is being considered to replace coal in aging plants. A fair comparison would be the best new gas plant to the most advanced coal plant because when the old coal plants are retired, the choice will be the best of the alternatives. A combined cycle natural gas plant has about 60 percent efficiency compared to about 43 percent for combined cycle supercritical combustion coal plants. This simply cannot be ignored and would skew the Howarth analysis toward gas even if their data were not suspect.

Howarth et al. defend their choice on the grounds that natural gas is used for purposes other than electricity generation (see commentary in the Michael Levi blog cited above). While this is true, coal is used predominantly for power, so in choosing to compare natural gas with coal, post-combustion analysis is appropriate.

Where the dueling reports diverge is in the area of fugitive emissions—releases of methane during the operations involved in producing and transporting the fuels. The quantities are in dispute, not the origins of the emissions. The Howarth data are from a variety of sources of variable quality. The most recent EPA estimates are much lower, as are those, predictably, from the gas industry.

Methane is about 25 times more potent than carbon dioxide in its global warming proclivity if the period under study is 100 years. The bulk of the debate surrounding the Howarth et al. work has been around the unconventional time scale of 20 years for the analysis, which most disadvantages methane. When carbon sequestration in deep saline aquifers is considered, the yardstick the practitioners are held to is well in excess of 100 years. In other words, the sequestered gas has to be guaranteed to not leak over that period.

In the case of coal, the emissions comprise methane found in association with the coal. For centuries this has been a known hazard of coal mining,

Global Warming Potential of Methane

Methane in the atmosphere will absorb infrared radiation attempting to escape to space. In so doing it causes atmospheric warming, much as does carbon dioxide, albeit at a lower level. Consequently it is classified as a greenhouse gas.

Over a 20-year timeframe, the global warming potential of methane is 72 times that of carbon dioxide. This means that if equal weights of the two gases are released into the atmosphere, methane will be 72 times as effective in warming. After about a 10- to 12-year period, the methane reacts with hydroxyl radicals in the atmosphere to produce carbon dioxide and water. This reaction product continues to warm the earth long after the methane has been consumed. Over a 100-year timeframe, methane has a net effect 25 times that of carbon dioxide, as compared to the factor of 72 in 20 years. The time period of comparison is therefore critical. Experts debate the appropriate timeframe, although current consensus leans to the higher number and is the number used by the Intergovernmental Panel on Climate Change (IPCC).

both from the standpoint of a poisonous atmosphere for miners and from the possibility of explosions in confined areas of the mines. In the past, canaries were famously used as indicators of methane. If they died, you got out in a hurry—a sort of go/no-go device. As previously mentioned, methane has no odor; commercial producers deliberately introduce an odor into methane for public use precisely for the detection of leaks.

Methane emissions from coal mining vary depending upon the nature of the mine. Deeper coal mines need to be purged for safety reasons, and therefore more methane is produced from them. The gas is purged with air, and then transported to the surface in vents. Consequently the methane is very dilute, usually less than 1 percent. This is the dominant source of methane from coal mines and is currently released to the atmosphere. Recently attention has been paid to capturing and utilizing the methane (Somers & Schultz, 2010). By many accounts much can be accomplished here economically. The next biggest source is when holes are drilled to purge the mines prior to actual operations with personnel. Mountaintop and other open pit mines likely have fewer emissions, but these are also harder to assess.

The Howarth study, flawed or not, has on balance been very good for all concerned. This is because it drew attention to the issue and drove a number of other studies that might not otherwise have been conducted. Such attention would likely not have been given were it not for the controversial comparisons with coal. With a large number of coal plants scheduled to be mothballed over the next few years, natural gas as a transitional fuel to more sustainable sources has great currency. Since that is the primary issue, Howarth's not taking into account the combustion step is a significant error. A direct consequence was the strident response, in the form of studies, from the likes of Worldwatch and the National Resources Defense Council.

The brouhaha certainly has focused attention on the fact that something concrete ought to be done about fugitive methane emissions. This ought to go beyond coal mining and oil and gas production. It should include the following two sources:

- According to the latest available EPA estimates, the third-largest US source of fugitive methane is livestock; in Europe and Canada it is the principal source. The rumen, or forestomach of animals classified as ruminants, converts feed by bacterial action known as enteric fermentation. One byproduct of this action is methane, which is expelled by the animal. The methane formation represents about a 7 percent loss of efficiency in food conversion. It seems the only truly viable approach to mitigating this is not capture, but amelioration. Modeling has demonstrated that up to 40 percent reduction may be possible by various interventions (Benchaar et al., 2001). A 2007 European study on dairy cows claims 27 percent to 37 percent reduction in methane through addition of just 6 percent lipids in the feed (Martin et al., 2007). For this study they used linseed lipids at a dairy farm. Since methane production represents inefficiency, cost-effective reduction is in the economic interests of the farmer.
- Methane from landfills is the second-largest source in the US, and some estimate it to be the largest. In the absence of sufficient oxygen, organic matter decomposes to a mix of gases, about half of which will be methane. Intervention can be in two forms. One would be to divert organic matters such as household food scraps and yard waste to a specialized facility.

Until such measures can be instituted, the alternative, capturing the methane from landfills, is feasible. This can be flared, which is burning at the end of a pipe (the resulting carbon dioxide being less harmful than methane), or it can be combusted for a purpose. This purpose could conceivably be generation of electricity, hot water, and the like. The business model of landfill operation does not necessarily fit this type of process.

Natural gas production and distribution can cause the leakage of methane in two principal areas. One is in transportation. In some cases the high-bleed design of pneumatic systems in the pipeline infrastructure is the culprit. Replacement with low-bleed designs is feasible, and substantial gains can be achieved by doing this. A higher-cost solution is replacing the gas with compressed air in the systems.

Also, the system of pipelines and associated valve assemblies can leak at various points after aging-induced malfunctions. But this can be addressed through maintenance mechanisms. One reason for the global drop in this sort of release is believed to be a Russian effort to replace old equipment.

The second and main source of fugitive emissions is the natural gas produced before a pipeline is in place to move it. This occurs in the early days of the prospect. Even in areas riddled with pipelines, a spur line to the new rig in question does not exist at the outset. Some operators may choose to not invest in a spur line until the reservoir is proven commercially viable. In that case, the initial gas produced during the discovery process has nowhere to go. It is often released.

A simple solution would be to flare it. This would dramatically reduce the problem since the released pollutant would be carbon dioxide, not methane. Howarth et al. assume that every bit of this gas on every well is vented. A 2012 study by Matthew Harrison of 1,578 shale gas wells across the country reports that in 93.5 percent of the cases, the spur line to the main gathering line is laid in time and the gas is recovered for sale. Thus only 6.5 percent of the gas in question is either vented or flared. This is a far cry from the 100 percent venting assumption of Howarth et al., so the situation is far less dire than they make out. The Harrison study was industry-sponsored, which could carry taint with some. But it is unlikely that the numbers are off by much. The larger operators such as Shell have publicly stated that they utilize all the gas. The most authoritative study to date, sponsored by EDF and conducted at the University of Texas at Austin, is ongoing, and the first report was published in late 2013 (Allen et al., 2013) in the *Proceedings of the National Academy of Sciences*. Their conclusions, if representative of the industry at large (theirs was necessarily a smaller sampling), are encouraging. They estimate that only 0.42 percent of gross gas production is released in some form. Some have challenged these numbers, but further reporting will hopefully allow a consensus on this issue.

The public may well ask why something useful is not done with the gas in every case other than simply putting it on the spur line, which, as we discussed is sometimes not present. The answer lies in part in the short duration of the production of natural gas wells. It cannot economically warrant any sort of capture and use. If such an economically enabling technology were to be developed, the potential to reduce methane emissions would be significant. The actual act of capture is straightforward and requires no particular innovation. The only issue is the economic utility of capturing gas as opposed to flaring. In some ways it is the same issue faced by landfill operators.

A little-known fact is that LNG is kept cold during the long voyage from Qatar or elsewhere by periodically releasing small quantities. This release causes chilling primarily through evaporative cooling. But in recognition of the economic value of the gas, it is collected and used in engines on the ship.

Natural gas associated with oil production when not in economically useful quantities is another potential source of emissions. Here, too, flaring is an option, as is any new technology to utilize the gas. Oil storage is another source, in that oil storage tanks are vented to release methane; in principle this methane could be recovered. In all recovery schemes, a yardstick for cost breakeven could well be the price set on carbon. Today there is no such thing in the US, but in Europe that price ranges from about $15 to $35 per tonne of carbon dioxide. One could use $30 per tonne as a target figure that one could reasonably project as an effective US-based "tax," no matter what the manner of implementation.

The US EPA issued the so-called green completions regulations (EPA, 2012) covering this issue. In summary, operators will be required to use measures to drastically reduce methane emissions during the early stages of production (the flowback period). They will have until 2015 to achieve this.

In my opinion, this is a reasonable policy. To expedite execution, the industry would be well served by sharing best practices with the smaller operators. This technology is unlikely to have a proprietary component, so sharing ought to be straightforward. Although research is not involved, RPSEA (the Research Partnership to Secure Energy for America) could be instructed to execute this aspect. This would be a fit with the small producer assistance stricture under which RPSEA operates.

The new rules also govern some of the other areas we have discussed in this chapter, including oil storage tanks and valve systems. There are also specific reporting requirements, which ought not to be onerous and are needed to assure a concerned public

A final note: Given that belching bovines are such a major part of the methane emissions equation, an outbreak of vegetarianism would help the environment!

CHAPTER 10

Earthquakes: Should We Be Concerned?

"Shake, shake, shake"

—From "(Shake, Shake, Shake) Shake Your Booty" by KC and the Sunshine Band (written by H.W. Casey and R. Finch)

Fracturing operations produce seismic energy. This can cause moderate earthquakes if the operations are proximal to active faults. Conventional techniques can identify such faults and interaction with them can be avoided. With diligent pursuit there ought to be no concern.

In Lancashire, United Kingdom, an earthquake of magnitude 2.5 on the Richter scale was recorded on April 1, 2011. Despite the date being subject to some tomfoolery, this actually did happen. It was followed by a 1.5-magnitude tremor on May 27 of the same year. A study by European scientists concluded that it very likely was tied to fracturing operations. It also mentioned that this was due to a combination of circumstances unlikely to be repeated and that the maximum such that could be expected was a magnitude-3.0 tremor.

In 2008 and 2009 the town of Cleburne, Texas, experienced a series of tremors up to 3.3 on the Richter scale. This town had no recorded history of earthquakes, so the residents speculated that fracturing operations, which were a relatively recent activity, were the cause. A team of scientists from two major Texas universities concluded that fracturing was not the likely cause but that waste fluid disposal wells could be implicated, and the two were related in that the waste fluid was mostly from fracturing operations.

In late 2010 and early 2011 a series of earthquakes were recorded in the Guy-Greenbrier area of central Arkansas. The US Geological Survey studied the phenomenon and dubbed it the "Guy earthquake swarm." Initial reports were not conclusive except to suggest that fracturing per se is an unlikely cause, but deep disposal wells could be the culprits. Of note, though, was the report that in Enola, southeast of Guy, swarms of about 3,000 quakes occurred in the early 1980s, and about 2,500 again in 2001, predating fracturing and deep disposal in both cases.

The Tohoku-oki earthquake off the eastern coast of Japan in 2011 was the fifth earthquake globally since 2004 of magnitude 8.5 or greater, an extraordinary recurrence rate considering that the last preceding earthquake of that magnitude was in 1965. The Tohoku-oki earthquake certainly cannot be compared with the others, in part because it could not possibly be human-induced except in a James Bond fantasy. But is the earth entering a period of increased seismicity?

The December 2011 issue of *Science* reports (Mueller & Yeston, 2011) on a paper by Andrew Michael in *Geophysical Research Letters*: "Recent seismicity can be described by random and high variability of low-rate events within a Poisson process rather than a cluster of related events." The main purpose of Michael's paper was modeling to show that the recent spate of large earthquakes, most lately the one mentioned above, was not an indication of an era of increased global seismicity. His point being taken, one could still conclude that higher incidences of small temblors were natural events and not anthropogenic.

But whether fracturing could develop enough energy to cause geological platelets to move deserves study. The same goes for stresses caused by deep disposal wells. But first some basics. Earthquakes are measured on the Richter scale, named after a Caltech professor who devised it. The information in the box at right comes from the USGS Earthquake Hazards Program website (http://earthquake.usgs.gov) and presents intensity in common terms. The Richter scale is logarithmic. This means that each increasing whole number indicates a 10× increase in displacement (ground motion) and about a 32× increase in energy. More recently, as shown in the box, the USGS has moved toward using the Modified Mercalli Intensity scale. Unlike the Richter scale, it has no mathematical basis. It simply is a ranking based upon the actual effects felt by the populace. USGS feels the public can better relate to such a scale. However, the popular press did not get the memo and still report these events on the old Richter scale. Some habits are hard to break. In an effort to be even-handed, I have provided both in the box.

The USGS table raises several points related to earthquakes observed in the vicinity of fracturing activities. The ones in the UK really ought not to even have been noticed. Perhaps an observatory recorded and reported it. The ones in Cleburne as well were barely in the noticeable range. But once noted, the recurrence would have raised concerns simply because they were not the norm. Not mentioned in the chronology above were some larger ones in Ohio,

Earthquake Intensities and Frequencies

Magnitude	Typical Maximum Modified Mercalli Intensity	Annual Average Number
1.0–2.9	I	1,300,000*
3.0–3.9	II–III	130,000*
4.0–4.9	IV–V	13,000*
5.0–5.9	VI–VII	1,319**
6.0–6.9	VII–IX	134**
7.0 and higher	VII or higher	15**

* estimated

** based on observations

Abbreviated Modified Mercalli Intensity Scale:

I Not felt except under very favorable conditions

II Felt on upper floor of buildings

III Vibrations similar to passing truck

IV Felt indoors noticeably, outdoors sometimes; like heavy truck striking wall

V Felt by nearly all, unstable objects topple

VI Felt by all, some frightened, damage slight

VII Poorly built buildings damaged, slight to moderate damage in well-designed normal buildings

Source: USGS, n.d.a and USGS, n.d.b

around 4.9, and an even larger one in Oklahoma, around 5.6, in late 2011. These appear definitely to be implicated with injection wells.

The other takeaway from the table is the sheer number of smaller events. The normal number worldwide in the intensity range of the suspect ones noted above is a reason to believe that many of those observed may be naturally occurring. Also, aside from a fear of the unknown, the moratorium on shale gas development called for in the UK based on those tiny quakes, the sort of which naturally occur a million times annually in the world, smacks of overreaction. This is especially so in a country reliant on imported gas. Single-issue activism without consideration of the other implications is always disappointing.

Seismic Activity Related to Fracturing

Many fracturing operations are closely monitored for seismic activity. This has been done for a number of years in order to map the pattern of fractures in an effort to improve drainage of the reservoir by making sure they knew where the fractures went. As a consequence, very detailed information is available. The magnitudes have always been known to be small, so the technique is designated "microseismic." In principle, monitoring wells could be placed in every new prospect to estimate the actual energy produced from fracturing in that rock. The most detailed studies to date are in the Barnett Shale. The levels observed are mostly under 3.0 in intensity.

The more important point is probably not the energy from the fracturing itself but rather the proximity to faults of reasonable size. According to the USGS, the magnitude of an earthquake is directly proportional to the length of the fault. 3-D seismic is a technique routinely used by the industry to delineate the subsurface. Essentially a three-dimensional picture of the reservoir and the rock surrounding it can be produced. This is routinely performed in offshore tracts but not so commonly on land, where a more cost-effective two-dimensional picture could do the trick. In prospective areas known to be significantly faulted, the prudent approach would be to run a 3-D survey to assist in placing the wells in the most productive areas while also avoiding faults of significant size. In areas with some risk, at least the initial fracturing operations may need to be accompanied by microseismic monitoring in real time. If a fault is intersected, the entry will be detected and the operation can be shut down.

Seismic Activity Related to Disposal Wells

Disposal wells have been more directly implicated in tremors. In Guy, Arkansas, four disposal wells were judged to be close to previously unknown faults. After two of these wells were closed, the incidence of earthquakes greater than 2.5 dropped significantly, and a moratorium was placed on wells in a 1,100-square-mile area over the fault system. Similar measures were taken in Ohio after additional quakes were linked to disposal wells.

There is a general belief that disposal wells, and these could include wells for carbon dioxide sequestration, have the potential to create seismic activity. While even this may not be of a damaging scale, it is disturbing for the public unused to such activity. Since disposal wells are regulated by the EPA, planning and monitoring ought to be straightforward. This in fact is likely why swift action to shut down was taken in Ohio: the wells were being monitored by the state Environmental Protection Agency. Careful planning and execution are critical for disposal wells.

These wells are used to dispose of all manner of liquid waste, not just that from drilling operations. Considering just the latter, one ought to give serious consideration to reuse of produced water being the primary option rather than the secondary.

Should We Be Worried?

In short, no. The seismic activity from the act of fracturing is small in intensity. It will by and large be below the threshold of human detection except in unusual situations. But care should be taken to not operate proximal to faults over a certain size. There ought to be rulemaking on this point. The technology of subsurface mapping to identify the location and size of faults is well known and currently used for practical reasons: operators do not want to lose valuable fluid down these faults. The technology of real-time monitoring of fracturing is well developed as well. Best practices ought to be developed with respect to the minimum number of wells in each prospect that ought to have such monitoring implemented. Good subsurface mapping combined with procedures to ensure no intersection of fractures with faults ought to be mandated.

Disposal wells do present a larger risk of detectable tremors, but again not likely large enough to cause damage. But the intersection of active faults by any such disposal wells ought to be avoided using very similar techniques to those mentioned above. This being an EPA-regulated activity, the execution of new measures ought to be straightforward. Strong consideration should be given to incentivize reuse of flowback water rather than deep disposal of it. Aside from avoiding the problem of possible tremors altogether, this would put less of a burden on fresh water withdrawals because at least a portion would be reused. In many, but not all, instances the cost may not even be greater.

PART III

Economics of Production and Use

III. Economics of Production and Use

The bulk of the "promise" element of the book is in this section. Reporters in reputable newspapers had questioned the very profitability of shale gas production. Growth in the industry bolsters the fact that the profits were there.

In the span of 2012 to 2015, shale oil has taken center stage away from shale gas. I had missed that trend in the first writing in part because shale oil burst into prominence in the very year of that writing (2011). It seems I was not the only one caught unawares. Pipeline infrastructure did not keep up with production, and in early 2015, a million barrels a day are being transported by under-regulated rail, with attendant issues such as explosive derailments. So, here we have an environmental impact issue not captured in Section II.

On the economic plus side, shale oil has caused a sustained drop in the world price of oil, a boon for net importing nations such as the US, China, and India. Over time this increase in domestic production of oil will affect politics as well.

Sustained low domestic prices have brought chemical processing companies, such as makers of ammonia fertilizer and methanol, to the US. The US will likely be a net exporter of these in the future. In addition, cheap natural gas is enabling the substitution of gasoline and diesel in transportation. Innovations in technology such as conversion of gas to liquids on a small distributed basis are discussed in this section. I construct a thesis whereby gas could be converted to liquids and sent down the Trans-Alaska Pipeline, providing a lifeline to a system in danger of extinction due to low oil volumes.

CHAPTER 11

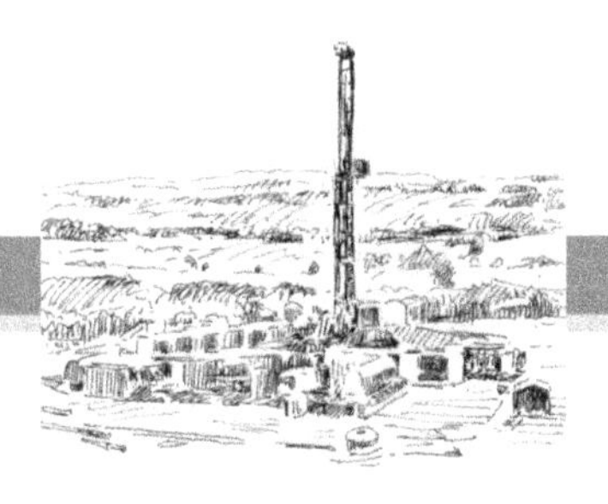

Is Shale Gas Production Indeed a Giant Ponzi Scheme?

"You're just too good to be true"
—From "Can't Take My Eyes Off You" by Frankie Valli (written by Bob Crewe and Bob Gaudio)

A *New York Times* top-of-the-fold front page piece, "Insiders Sound an Alarm Amid Natural Gas Rush" (Urbina, 2011), discussed the profitability of shale gas wells and is very bearish on the prospects. The author is careful to use the term "Ponzi scheme" in a statement attributed to someone else. But as anybody with modest discernment knows, such a reference made in the early stages of a piece is at the very least a tacit endorsement.

I acknowledge the principal points: some in the industry worry about profitability, especially given the low gas prices in the last year or two. I present here a case for rejecting the pessimistic premise. These are early days in the exploitation of a completely new type of reservoir. Continuous improvement, as in any industrial endeavor, can be expected. In the case of shale gas, the learning curve is likely to be steep. In part this is because of the sheer volume of activity. Each well can be drilled and produced in as few as 21 days, so the financial risk is low and the number of wells will be large. The setting is almost akin to a factory, the type of enterprise amenable to rapid learning curves.

Production From Shale Gas Wells Declines Rapidly

True. The decline is steep, with a drop of 60 percent to 80 percent in the first year. (Conventional reservoirs decline 25 percent to 40 percent in the first year.) After year two, there is a gradual asymptotic decline. The mechanism is still being debated, but premature closure of the fractures is a likely explanation. This could be due to insufficient penetration of proppant into the formation. Industry is working on materials and techniques to cause improved and more sustained flow. A Rice University–originated product sourced from nanomaterial is in early stages of commercialization. This and other such products are being designed to be lighter than conventional proppants, while still being strong enough. Being lighter they can be expected to float out further into the cracks.

Refracturing

In this technique, new fractures are initiated in existing well bores, often directly on top of the old ones. In the few cases where it has been attempted in the Barnett, the results have been dramatic. Production rates after refracturing have reached and exceeded the original starting production. And sometimes they decline at the same rate as before. This is indicative of the possibility that new rock pores are being accessed.

Current research at the University of Texas indicates that the optimal time to refracture is two to three years after initial production (Sharma, 2010). The University of Texas study will also examine other factors such as precise location relative to the old fractures. One hypothesis is that closure of the fractures in the zone produced induces a stressed region, which discourages new fractures from going to that area. In some observations, the new fractures are seen to bend away from the depleted zones.

Somewhat ironically, a shortcoming of the resource, poor permeability (a measure of the ability of fluids to flow in the rock), may be why refracturing works. Ordinarily, poor permeability means less flow, and hence less production. Fracturing improves that. But if the fracture paths are impaired, as explained above, the gas does not get fully drained from adjacent rock. However, it remains available for new fractures, and is for all practical purposes from new rock despite being proximal. From the standpoint of economics of the prospect, all that matters is that each operation cause enough production to ensure a rate of return. The fast declines are not highly material if this economic threshold is met. One final point: refracturing comes at a fraction of the cost of the original well because no new well bore is drilled. So the newer gas has a cost basis that could be a third or less of the initial gas. This does wonders for prospect economics.

Wet Gas

There is a passing allusion to wet gas in the *New York Times* piece, but it deserves serious attention because of its dramatic effect on profitability. Wet gas is defined as natural gas with a significant component of hydrocarbon species other than methane, known collectively as *natural gas liquids* (NGLs). The principal constituents are ethane, propane, butane, and even larger molecules broadly named *condensate*. As a detail, although ethane is lumped in with NGLs, it is actually a gas at ambient temperature and importantly also a gas at temperatures that condense out the bigger molecules propane and butane. This distinction is important in the method of separation from methane and is discussed in a separate chapter.

The economic significance of NGLs lies in the spread between natural gas and oil prices. Natural gas, on the basis of energy content, is currently priced at about a fourth of oil. Decades ago their prices were in parity. Natural gas liquids, the "wet" part of wet gas, are priced in relationship to the price of oil. Condensate is at or somewhat higher than the price of oil, and butane is definitely higher than oil because it is essentially a drop-in replacement for gasoline. Propane is at a discount to oil, as is ethane. Ethane is the least costly, at about half the price of oil. The actual price varies depending on location and availability. But all these are vast improvements over the price of methane which is typically one-third to one-fourth the price of oil.

A typical Marcellus wet gas is reported by one oil company as pricing out about 70 percent over dry gas. Range Resources reports that at a flat $4 per MM Btu gas price (incidentally, the average for 2010 was around this figure, and in 2013 that was close to the average as well), its internal rate of return would be 60 percent. That is way more profitable than many conventional gas prospects. But you and I can do our own calculations (see box below)—or you can skip that and go to the punch line on returns.

Wet Gas Economics

The wet portions of North American shale gas deposits average between 4 and 12 gallons of NGL per thousand cubic feet (mcf) of natural gas. Typical Marcellus wells run about 1,500 mcf per day, and an average cluster of wells (pad) may have 15 wells, giving daily production of 22,500 mcf per day. Using a figure of 7 gallons NGL per mcf, that yields 157.5 gallons, or 3.75 barrels of NGL per day.

Ethane tends to average 60 percent of the NGL. (The full implications of the lowest value NGL being so preponderant are discussed in chapter 14, "The Ethane Dilemma.") For simplicity I will count all the other liquids priced at a discount to $100 per barrel of oil, at $70 a barrel. Support for this is the EIA-sourced pricing figure in chapter 14. That figure demonstrates the recent history of NGLs at roughly 80 percent of crude price and, separately, ethane at about 50 percent of crude price. In both the NGL and the ethane, I have been more conservative than that. Ethane I will price at half of that, at $35 a barrel. Ethane prices out at 0.6 x 3.75 x $35 = $78.75. The other NGLs collectively are worth: 0.4 x 3.75 x $70 = $105. Total value of the NGLs is the sum of those two: $183.75. (Note that I call ethane an NGL, following industry practice, but in the calculation I make a distinction because the EIA figure splits out ethane.)

To estimate the effect of NGLs, it is simpler to reduce the above figure to that associated with 1 mcf gas. That would be $183.75/22.5 = $8.17. This calculation states that the NGL associated with natural gas could add $8 to the typical price of natural gas these days, which was $4 per mcf in 2011. To be immensely conservative, let's halve the NGL value by discounting severely. That still indicates double the profits over dry gas. The actual market value of such liquids can vary by area, which is why I have chosen to be so conservative. But the moral is that the wet component dominates profitability.

Wet gas profitability is shown to be more than double that of dry gas. A downside would be drops in the price of oil, thus reducing the NGL value. In chapter 2, "The Oil Plateau and the Precipice Beyond," I describe models in support of the belief that oil prices will remain high except for the usual perturbations driven by external factors. Another downside is wetness at the lower end of the scale mentioned, below the average figure of 7 gallons of NGL per thousand cubic feet (mcf) of gas. Keep in mind, though, the heavy discounting we did in our calculations relative to oil. Even accounting for the costs to clean up the liquids and transport them, the value of NGLs should be closer to the price of oil than our conservative assumptions.

The Marcellus Shale, the largest and most prolific of the North American deposits, has a wet character on its western side. The as-yet not important producing states of West Virginia and Ohio are advantaged in this regard, as is western Pennsylvania. The Utica Shale is described in chapter 4. It is a newly discovered province that promises to be bigger and more productive than the Marcellus, but these are early days yet to put any certainty on the size. Its productivity, on the other hand, is more of a sure thing because the Utica is set deeper and so can be expected to have higher natural pressure to drive the fluids up. On average it has wetter character than the Marcellus, again on the western side. The same three states are advantaged by this.

How Things Will Play Out

Given the facts above, expect the wet gas prospects to be produced first. Over the next few years, the price of dry natural gas will rise because of demand. Massive switching from coal-fired electricity to gas will occur. This is because even without a price on carbon, the all-in cost of electricity from gas is less than that from coal at gas prices below $8 per MM Btu. In a 2011 article (Rao, 2011), I presented a model predicting gas prices to have a lid at about $8. This stability will contribute to switching from oil to gas. The switches will include methane propulsion of vehicles and gas-to-liquids-derived diesel and gasoline. Chemicals traditionally derived from oil will switch to being produced from natural gas or the associated NGLs. These chemicals include ethylene and all its derivatives and propylene. Many of these products are imported today.

Over time, all this oil replacement, plus electric vehicles, will make a significant dent in our $300 billion plus annual imported oil bill, and hence our balance of payments. Importantly, gas prices will be less subject to the whims of the weather because heating and cooling will be an ever-decreasing component of gas usage. Also, with shale gas production heavily distributed and deep inland, hurricane-related disruptions to gas supply, and associate price spikes, will be a minor factor. But 33 of the 42 US/Canadian ethylene crackers are on the Gulf Coast. So hurricane-related reduction in the supply of chemicals that are based on natural gas is still not out of the question.

Demand creation will enable a gradual return to dry gas production. Some of the earlier plays are profitable at $4 already. But a rise in the floor price will ensure the supply that will be available when the consumption trends described above mature.

And one day *The New York Times* will have a page 1 above-the-fold piece on how shale gas transformed the US economy. Then I will wake up.

CHAPTER 12

The Supporting Actress Is Now the Star?

"Lightning is striking again"

—Lou Christie in "Lightnin' Strikes," written by Lou Christie and Twyla Herbert

Just as we were absorbing the enormity of the economic impact of shale gas, along came shale oil and in many ways trumped it. Oil used to be in the supporting cast for shale gas. Not oil per se, but natural gas liquids (NGLs) found in association with the gas, had value. Economists tend to lump NGLs with oil in their statistics. I discussed in chapter 11 the fact that the NGL component was the difference maker in the economics of the shale gas production as long as natural gas prices stayed low. This is driven by the fact that most NGLs are pegged to the price of oil rather than gas and oil continues to be four to five times as costly as natural gas.

Supporting Actors Stealing the Show

It does not happen often, but when it does, a star is usually born. Abigail Breslin saved the otherwise inconsequential movie *Little Miss Sunshine*; it's too early to see what the more grown up Breslin will do. Austrian actor Christoph Waltz stole the show from Brad Pitt in *Inglourious Basterds* and got an Oscar and out-of-nowhere recognition for his work. These performances basically demonstrate the actors' credo: there are no small parts. In opera that is often literally the case. When the novice Frederica von Stade came out of obscurity in 1973 in the role of the page boy Cherubino in *The Marriage of Figaro* first in Santa Fe and then at the Met in New York (I was privileged to be there), she did steal the show. But was it in fact a small part? Cherubino it could be argued has two full arias, same as the Countess. But when was the last time you saw the person playing Cherubino advertised on the poster? Von Stade, known as Flicka to her adoring legion of fans, went on to be a star.

Shale Oil

The pricing disparity has caused a radical shift toward shale oil away from gas, especially "dry gas," which is natural gas with little or no NGL content. There are no hard numbers for this but if the NGLs are less than about 2 gallons per mcf gas, then the gas could be considered dry. Typical producing wet gas wells run between 4 and 12 gallons per mcf.

The logical extension to wet gas is oil itself. In chapter 1, I explained that from the standpoint of chemistry and geology, oil and gas are one vast continuum. Consequently, where you found one, the other was also likely to be found. Probably the fastest growing shale hydrocarbon area in the US has been the Eagle Ford play in Texas. This was driven in no mean part by the fact that the single area contained fluids ranging from dry gas to oil. Consequently, property owners became very conversant with the special operational requirements of producing oil, some of which are described later in this chapter.

The big story in shale oil is the Bakken formation. It lies mostly in North Dakota but also is found in Montana, South Dakota, and Saskatchewan. The existence of recoverable oil was known for a couple of decades, but serious exploitation did not happen until about 2008. As one can see from Figure 7, the explosive growth took a couple more years, and I missed noting it in the first edition of this book, which I wrote in 2011. The pace is keeping up, and shale oil produced from the Bakken is right around 1.3 MM bpd in early 2015.

Figure 7. Bakken and Eagle Ford oil production history

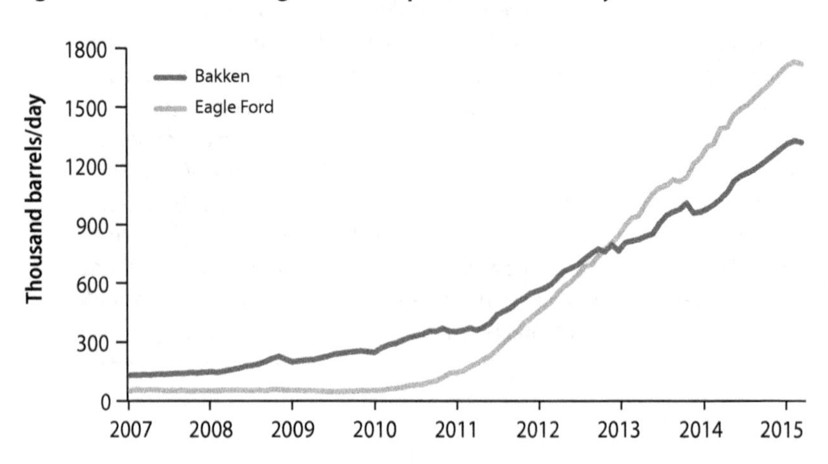

Source: US Energy Information Administration: Drilling Productivity Report

The Eagle Ford, as I mentioned, has both gas and oil. In 2011, attention shifted to oil production, and the growth rate (shown in Figure 7) is even more remarkable than that of the Bakken. In part this high rate is because being in Texas, the Eagle Ford formation is close to petroleum industry infrastructure and potential markets for its product. Even so, marketability of this light oil has been challenging, as I discuss below. Not surprisingly, the pace of development pickup is similar for the Bakken and Eagle Ford plays.

As of early 2015, together these two sources of oil are producing over 3 MM bpd and increasing. Other plays such as the Niobrara (300,000 bpd) and the western Marcellus are also adding material quantities. This is against a backdrop of oil *imports* that were over 8 MM bpd a scant few years ago.

All of this is causing bold, possibly rash, predictions of oil independence. This domestic source arrived contemporaneously with reduced utilization, especially of gasoline, through higher CAFE standards and the unconnected recession we experienced. In chapter 26, I discuss the further reduction in domestic oil demand through the use of natural gas as a raw material for a number of fuels and chemicals ordinarily produced from oil. In fact, I believe the world will see a progressively reduced dependence on oil. This will begin with gas substitution mentioned above and then later be joined by biomass-sourced liquids and chemicals. This will happen neither quickly nor completely. In other words, oil is here to stay for a long time. But it will lose its status as an essential or strategic commodity and simply become a useful option. When that happens the price will moderate. Nevertheless, energy *independence* is not a particularly viable concept in this connected world. But energy *security* is a worthy objective.

The Nature of Shale Oil and the Economic Consequences

Shale oil is a throwback. This is the way oil used to be back in the day, way back. We are talking over half a century ago. Industry explores for oil and gas and tends to produce the easy stuff first because that delivers the most profit. The result is that conventional oil is now by and large either not easy to produce or not as desirable when produced. In the first category is oil from ultra-deep water (over 5,000 feet water depth) and demanding environments such as the Arctic. In the second category, the less desirable oil, in conventional settings, is defined by its properties. Typically it is heavier or sourer or both. Heavy means a high proportion of large molecules, making transportation and refining more difficult and expensive. Sour means more sulfur in the oil or gas. In the case of gas, the less desirable category includes a high proportion of carbon dioxide. Very few prospects are being exploited with more than 10 percent. The bypassed gas can have as much as 60 percent carbon dioxide. Might as well call it a CO_2 well, not a natural gas well!

In some ways the shale oil and gas revolution has turned the industry on its head. Sweet gas (defined as not sour) is now available from shale and other tight rock. A defining feature of shale gas is that it is almost exclusively sweet. Similarly, shale oil is almost universally sweet, and it is light to boot. Consequently refiners should welcome it with open arms, right? Wrong. But before we get to that particular travesty, consider that shale oil and gas are throwbacks to the good old days because they are a brand new resource. When conventional resources were new, there was a lot of the light sweet fluid. The other reason is geology: shale oil and shale gas are in the source rock, the place they originally formed. When either formed it was light and sweet. Virtually all the oil that became heavy or sour got to that state due to bacterial action, usually at shallower depths (all the important heavy oil deposits are relatively shallow).

Domestic Market Is Weak for Shale Oil

The vast majority of US refineries are of the complex variety. Many have invested heavily in the specialized equipment to handle heavy crude from Canada, Venezuela, and Mexico. Some estimates have that investment at $20 billion (Phillips, 2013). Canadian crude imports in particular are on the rise. This extra-heavy crude is heavily discounted with respect to the West Texas Intermediate (WTI) price on light sweet crude oil in recognition of the refining costs and also because a portion of it, after the best refining efforts, has essentially no value. This final residue is known as petroleum coke and does have fuel value comparable to the better grades of coal, but it has limited use in this regard.

Light sweet shale oil ought to sell for WTI price. But US refiners essentially cannot afford to pay that price because it would underutilize their special equipment. Furthermore, US laws do not allow export of the light oil except to NAFTA countries. The net result is that this valuable oil has been selling at a discount compared with WTI. While the prices of the two benchmark crudes, WTI and Brent, tend to vary, Bakken crude at the wellhead in January 2014 was selling for $18 less than WTI and $27 less than Brent. That $18 spread with respect to WTI is not that far off the Canadian heavy (known as Western Canada Select, or WCS) spread, which in recent years has ranged from $15 to $25, with minor excursions higher.

Refineries and Refining Margins

Refineries fall into two categories: simple and complex. Simple refineries heat up the crude oil and separate the vapors into condensates of value, primarily gasoline, kerosene/jet fuel, and diesel. What is left over is known as fuel oil, which has uses such as for heating homes. The lighter the oil, the lower the fuel oil fraction. Simple refineries may also do some hydro-skimming, which is using hydrogen to remove sulfur. But they make no effort to upgrade the heavier molecules.

Complex refineries use a process known as cracking, which reduces the size of the molecules in fuel oil to create gasoline or diesel. This is done if the market for the transport fuels is good and better than that for fuel oil. The most common process used is known as fluidized catalytic cracking (FCC). The heavier the oil, the greater the need for cracking. Extra-heavy oil is often vacuum distilled, after which it is left as a residuum that has no use except as a partial blend with fuel oil. The residuum is usually processed in cokers, where intense heat breaks down the molecules. The end result is a solid known as petroleum coke, which cannot be further refined into gasoline or other fuels. The heavier the oil, the greater the fraction of this substance with limited utility. For this reason, extra-heavy oil is almost always priced lower than lighter oil.

Refining margins are calculated as the difference between the total selling price of the output minus the cost of the crude. What makes refining a less-than-predictable business is that both ends of that chain can have volatility. Simple refineries generally have lower margins unless there is an unusual depression in price of light oil. This is in fact happening with shale oil—more on that elsewhere in this chapter. Complex refineries have more costly equipment, but the feed cost is generally lower and more of the feed is converted to valuable fuel.

This is why the producers are lobbying for permission to export crude—because they are likely to get a world price for light sweet crude, at least WTI and possibly close to Brent. Pushback is coming from many fronts but in particular from the independent refiners (refiners not tied to an oil producing company). Given that refined products can be exported at will, the refiners are in a bit of a bind philosophically. They are currently enjoying the free market–enabled margins while arguing against the same for producers. In fact, the US currently exports over 3.5 MM bpd of refined products, as compared with an expected crude oil import level of 6 MM bpd in 2016. In general it is good for the nation to be exporting refined products because the margin stays in-country as a profit. Also, the creation of these products results in jobs.

The other interesting twist is that the artificially low price for light sweet crude has emboldened the resurrection of old simple refineries and even the building of new ones. Most of these are small—under 20,000 bpd and as small as 10,000 bpd. By comparison, conventional refineries are mostly larger than 100,000 bpd, and newer world-class ones such as the one in Jamnagar, India, are over 1 MM bpd. The new one is a 20,000-bpd plant near Dickinson, North Dakota, the first new refinery permitted and built in the US since 2008. No new refinery larger than 50,000 bpd has been built since 1977. Given the uncertainties with regard to permission to export crude, these simple refineries would do well to be small, thus reducing supply risk.

The Midstream Was Unprepared

The midstream is defined as the capability that transports produced fluids from the "upstream" well locations to the "downstream" refineries. Shale gas, and particularly shale oil, suffers from inadequate capacity to take it to market. A glance at Figures 7 earlier in this chapter provides a clue as to the reason. The rapidity of the buildup would have tested any midstream capability. The hardest hit area was the Bakken, where no prior capability existed. The result is that the oil is being transported in trains (and trucks to get it to the train).

A scant five years ago such a proposal would have been met with derision, yet these days trains are carrying nearly 1 MM bpd of North Dakota oil. The same is happening with Canadian heavy crude, especially in the face of seemingly interminable delays in the sanctioning of the Keystone XL pipeline. Incidentally, this pipeline was designed to carry a portion of Bakken oil as well, in addition to the Canadian crude. A major oil train derailment in Quebec in 2013, resulting in 47 deaths and significant environmental damage, gives one pause regarding this mode of transportation. At least two derailments of Bakken oil trains occurred in 2013 as well. Saying no to a pipeline, whether it is the Keystone XL or any other, is now an implicit acceptance of transport in trucks and trains. Opposition to pipelines in the past did not present that alternative. Consequently, activism in this regard must recognize the change in the landscape and decide which alternative is worse.

The absence of natural gas pipelines is leading to massive flaring of associated gas. This is natural gas that is found incidental to the oil production. In some ways this is an analog to NGLs found in association with natural gas. In this case, however, the gas has little value unless it can be taken to market. Release to the atmosphere being especially harmful, as discussed in chapter 9, this gas is flared. By some estimates North Dakota is flaring about $100 million

worth of gas each month, and this tally will continue to rise with more oil production. Many remedies are being sought, including building ammonia fertilizer plants in close proximity and making plans for pipelines. Some of the ideas in chapter 18, "Horses for Courses: Challenging the Orthodoxy in Fuels and Chemicals," could be employed here. But as it stands, the suddenness of the gas production has caught the industry unprepared and much profit is literally going up in smoke. Royalty owners are unhappy, and lawsuits could be forthcoming.

Implications to Energy Security

US oil production has exceeded imports for the first time since 1995 (US EIA, January 23, 2012). This is largely attributed to the surge in shale oil production. The EIA expects a continual rise in this production out to at least 2020. But this increasing self-sufficiency will not insulate the country from price shocks from world events. This is because oil is a world commodity, and a world market price exists. However, by importing less oil, more is available for other markets, thus damping volatility induced by singular events. Because shale oil is light and sweet, the US will soon not be importing similar oils from Nigeria and Angola. That oil will now be welcome elsewhere, especially in countries with predominantly simple refineries, which are especially suited to refining light oil.

Eventually the combination of higher domestic oil production and the conversion of abundant and low-cost natural gas to transport liquids will severely limit oil imports. By 2025 the new CAFE standards will also serve to reduce consumption of transport fuel. Not out of the question is the scenario that by 2025 the US will source all its oil from North America. While something short of oil independence, this would constitute a high measure of security of supply.

CHAPTER 13

Chemical Industry Prodigals Can Return

"Get back to where you once belonged"
—From "Get Back" by The Beatles (written by John Lennon and Paul McCartney)

Consistently cheap shale gas and associated liquids will transform the US chemical industry. Production that fled these shores in the face of high and uncertain natural gas prices will return.

Unrest in the Middle East causes spikes in the price of oil, with immediate impact on gasoline price. Rather uncharacteristically, the Islamic State violence in Syria and Iraq have not had the usual upward impact on oil prices. This is likely due to the balance brought by new US production. But even when oil spikes, the price of natural gas tends to remain stable over the same period. This story line has repeated often and underlines the principal difference in these two essential fuels. Oil is a world commodity, while gas is regional. Also they serve largely different segments of end use. Consequently, the fact that gas is one-half to one-fourth the price of oil (in energy content) has little relevance in the main.

However, if industry believes that this differential will hold for a long time, technology-enabled switching will occur. In this chapter I predict a shale gas–enabled future of gas at low to moderate price for a long time. At the same time I subscribe to the view of an upcoming plateau in oil production, which will drive oil prices higher. These two trends taken together assure a high oil-to-gas price ratio. The EIA 2013 forecast is in complete consonance with this view. This will cause systematic switching where possible. Industries that went abroad due to pricing uncertainty will gradually return. This will be good news for US balance of payments and job creation.

Predictably cheap natural gas stimulates a number of different avenues of endeavor. To begin with, a number of chemical and metallurgical industries have a high component of their cost tied up in fuel. The ones most affected are those with natural gas as feedstock. Also impacted are those in which oil can be switched to gas. One such industry, the production of ethylene from ethane rather than from oil-derived naphtha, is discussed in chapter 14. That particular compelling value move had an unintended consequence. Propylene production was impacted.

The Propylene Story

One of the derivatives of propylene, polypropylene, is ubiquitous in our lives: roofing, carpets, bottles, and bendable plastics, to name a few. For years, when oil and gas pricing was in greater parity, propylene was a byproduct of ethylene production in oil refineries. It is also produced by tweaking the catalytic cracking process, at the cost of a smaller gasoline fraction. A refinery can change the mix essentially at will, presumably based on the relative profit potential.

But with a worsening oil:gas price ratio, ethylene production increasingly switched to an ethane feedstock. Unfortunately this process produces very little propylene as a byproduct. So, as reported in *The Economist* ("Plastics Prices," 2011), in the two years preceding 2011 the price of propylene went up 150 percent and continued to stay at those high levels through 2013. Propane associated with shale gas production is easily converted to propylene by dehydrogenation, much as ethane is cracked to ethylene. But propane is a high-value commodity and is priced near oil, so this is a costly approach. A predictably low price for gas will allow for plants dedicated to producing propylene from gas. At least three companies, Lurgi, Total, and UOP, have the technology at an advanced state. Just as in the case of ethylene production, domestic locations for these plants are very likely.

Ethylene

This is the dominant feedstock for the plastics and fiber market. When natural gas prices started fluctuating in the early part of the last decade, much of the cracking capacity was shut down and the likes of Dow invested in plants in the Middle East and Asia, proximal to low-cost feedstock. The US became a net importer of ethylene derivatives. Now the US and Canada have predictably low prices compared to most of the consuming nations. All the major players have announced reopening of old plants and the construction of new ones. This is driven by two factors. One is the virtual certainty of high-volume ethane availability at low prices, and the other is the predicted demand in the coming years.

In a few years, the ethane supply from the shale gas–related NGLs will exceed the cracker capacity. As discussed in chapter 14, the newly built crackers could be in the Marcellus area, either as massive plants or as distributed small-footprint reactors. To get a feeling for how many crackers would be supported, I've done a calculation (see box below) using typical figures for NGL content of Marcellus gas. The size of crackers being discussed today by companies seriously considering building these is 1 million metric tons (tonnes) per year.

I conclude that such a unit would be served by about 420 wells. Pads with multiple wells will be increasingly common. So the number of pad locations serving each cracker will be somewhere around 20 to 30.

Crackers will take a few years to get into production. Assuming an immediate desire, we need to look at 2014 to 2015 targets. Some estimates put Marcellus production in 2014 to be between 5 and 10 billion barrels per day (bbd), of which a quarter may be wet. So, assuming a figure of 7.6 bbd, the wet gas will be 1.9 bbd, yielding 317,000 bpd of NGL, of which about 190,000 bpd will be ethane. That will cover three 1-million-tonne-per-year ethylene crackers.

Marcellus Assumptions

Typical values are presented below for the liquid-rich segments of the Marcellus Shale.

- One thousand cubic feet (1 mcf) of Marcellus shale gas would produce approximately 7 gallons of NGLs (the range is 4 to 12 in wet deposits across the country), ~60 percent of which would be ethane.
- Each Marcellus well produces approximately 1.5 million cubic feet (1,500 mcf) per day of wet gas. (Utica shale would be higher, maybe double, in part due to the higher pressure at greater depth.)
- Each well pad would contain around 15 individual wells on average. Pad drilling occurs only after full development is planned. Initially individual wells may be drilled. The economic, environmental, and societal benefits of pad operations will make them routine in most instances.
- Each well pad could thus produce around 94,500 gallons per day, or 2,250 barrels per day (bpd), of ethane (7 gallons x 0.6 x 1,500 x 15 wells = 94,500 gallons).
- A new 1-million-tonne-per-year ethylene cracker would need 63,000 bpd, or 1.3 million tonnes per year of ethane feed (Seddon, 2010). Thus, a single conventional cracker could require the ethane output of 28 well pads, or 420 individual wells.
- A 63,000 bpd natural gas processing plant would be expected to have a capital cost of ~$700 million, or ~$11,100 per bpd ethane, or ~$540 per annual tonne of ethane.
- Annual operating costs of such a facility would be on the order of $35 million (Seddon, 2010), or ~$1.50 per barrel, or ~$27 per tonne of ethane.

Financing for crackers always requires guarantees on long-term supply. If dry gas prices remain low, the proportion of shale gas exploited that is wet will remain high simply to assure more profitability. So, there would be a logical basis for security of supply. Furthermore there is an expectation that the low-cost driver will provide good margins for export of the ethylene or its downstream derivatives such as polyethylene, PVC, and polystyrene.

Nitrogen-Based Fertilizers

Modern agriculture relies dominantly on synthetic fertilizers. The most important one is ammonium nitrate, which also has an unfortunate use in explosives as well (think Oklahoma City bombing). Another is urea, much of which is used in the production of rice. As a major producer of crops, the US is a significant user of synthetic fertilizer. Much as in the case of oil, we use a quantity disproportionate to our population: 12 percent of the world's usage as against 5 percent of the world's population. The primary feed for this fertilizer is ammonia, which in turn is completely dependent on natural gas, which accounts for 90 percent of the cost. The high and erratic price of natural gas caused over half of the industry to flee to other parts of the world with low-cost gas. Trinidad and Tobago is the largest supplier by far, followed by Canada and Russia.

Cheap shale gas is luring this industry back. Given the importance of food to the nation, it would not be much of a stretch to suggest that fertilizer is a strategic commodity and that domestic production is a welcome change. A number of ammonia fertilizer plants have been announced for build by 2016. Prior to the flight abroad, the US was a net exporter of fertilizer. This could happen again. It might also not be off base to suggest the possibility of reduced food prices due to a consistently low fertilizer price. Were this to happen, the irony would not be lost that the last time the nation discussed the food/fuel nexus, it was the anxiety occasioned by beef prices rising due to the diversion of corn to ethanol.

Who Will Do It and Where Will the Jobs Be?

The chemical industry is vertically integrated by and large, in one direction or the other. Vertical integration involves being the producer of the feedstock as well. So, the major oil and gas refiners are also producers of the oil and gas. There are exceptions, such as Marathon, which is not a major producer of the fluids but is a huge refiner. Since production, known as the upstream, and refining, referred to as the downstream, are such distinct competencies, the decision to vertically integrate is a business one.

Oil field service companies perform the bulk of the work in the extraction of the fluids from the earth. With one singular exception, no oil field service companies participate in the processing. Even that exception went away with the divestiture of KBR by Halliburton a few years ago. A separate set of companies designs and builds refineries, crackers, LNG plants, and all other manner of chemical processing plants. In other words, vertical integration is uncommon in all aspects of the oil and gas service sector.

Oil companies have tended to vertically integrate from the very beginning. The big ones take it all the way from upstream to downstream to products and retailing. This is why you see consumer automotive products such as lubricants made by the likes of Shell. They are also into retailing, although many gas stations are franchised and not owned by the major oil company. As a curious point, although seemingly integrated vertically, the gasoline at an Exxon station may not have come from an Exxon refinery and almost certainly not from an Exxon-owned upstream facility. This is for two reasons. One is that refineries, no matter who owns them, will take feed from wherever they can get it. Refineries are finely tuned to accept certain input compositions, so the precise nature of the feedstock is more important than the source. The second is that gasoline distribution is most effectively accomplished using blending sites that are agnostic regarding the source, as explained in the box below.

BP Gasoline May Not Be From a BP Refinery

Gasoline blenders and distributors are located strategically to serve various filling stations. The fuel may come from any sort of refinery, and standardization has rendered this possible. The tanker trucks from the name brand stations drive up to the distributor and accept a product with a known octane rating. Blenders also put in the needed ethanol. Incidentally, the infamous 50-cent-per-gallon ethanol subsidy was being awarded to the blender, not the farmer making the corn. At this point the truck will blend in the special chemicals that each company believes impart all the qualities advertised at the filling stations: better engine cleaning properties and so forth.

In one sense, therefore, the boycotting of BP stations was misdirected, although the message value might have been present. The station being boycotted was most likely privately owned by a guy who was about as far from the Macondo oil spill as was a shoe salesman from the potentially questionable labor practices of a shoe manufacturer. The gasoline in question may not have any BP provenance other than the blending additives.

Another type of vertical integration is backward from the consumer product. In this case, the giants who make containers, fabrics, and all manner of plastic goods integrate back to make the key ingredient.

In all the discussion regarding the return of the chemical industry, the question arises as to where these jobs will end up. Certainly some of the jobs will be in the same locations as plants mothballed by the shift abroad. In this category falls the announced reopening of ethylene crackers in the Gulf Coast by Dow Chemical. But the new plants could well go closer to the source of ethane, as discussed in chapter 14. Bayer, Shell, and Total have expressed interest in crackers in the Marcellus area.

The domestic fertilizer growth will follow similar lines. Much of the current capacity is in the Midwest and the Gulf, and one could expect expansions to current plants. New plants, especially if dedicated to export, could go near the East Coast. For a bulk commodity such as ammonium nitrate, which also has transportation safety issues, proximity to end-use states could drive new plants to be in the breadbasket.

When a single commodity is a dominant portion of the cost of a product, comings and goings of plant locations are more to be expected than would be for processes with multiple commodities contributing to cost. Predictably low natural gas prices ought to persuade prospective owners of plants of a highly predictable long-term future. Such predictability keeps discount rates down and is a magnet for investor dollars.

CHAPTER 14

The Ethane Dilemma

"Why don't you stay?"

—From "Stay" by Sugarland (written by Jennifer Nettles)

The abundance of shale gas and the as yet limited demand have kept the price of natural gas low in North America. The price per million Btu (MM Btu) hovered at or under $4 for most of 2011 and hit decadal lows in early 2012. As conjectured in chapter 11, this is driving the activity toward wet gas production. In the principal producing areas more than half of the NGLs in wet gas comprise ethane. This abundance of ethane is posing the industry with a dilemma with respect to the most appropriate manner of exploitation. A good problem to have, but it is a dilemma nevertheless.

Ethane is a molecule with one more carbon atom than methane has. This crucial additional carbon allows for easy conversion to ethylene, which is ethane with two hydrogen atoms removed. Ethylene in turn is the basic building block of a host of useful products (see box on the following page). But ethane really has no other use except for increasing the value of natural gas. Natural gas nominally has 1 MM Btu per thousand cubic feet (mcf). But some sources, especially coal bed methane, can be as lean as 800,000 Btu per mcf. In these cases, ethane, with higher energy content, can be used to raise the value to a million or a bit more.

The processing of ethane to ethylene is done in plants known as crackers. The ethane dilemma centers on the fact that the crackers are located on the Gulf Coast, over a thousand miles from the workhorse gas production areas of Marcellus and Utica. To compound the problem, the majority of the ethylene-consuming factories making consumer and industrial products are located, you guessed it, right near the Marcellus and Utica fields. So one scenario would have the industry transport the ethane down to the Gulf Coast, convert it to ethylene, and ship ethylene derivatives all the way back again. This chapter is devoted to discussion of this proposition and alternatives, together with the technical and economic hurdles of each option.

Making and Using Ethylene

Ethylene has the formula C_2H_4. It is synthesized by stripping two hydrogen atoms off ethane, which is C_2H_6. This process is known as *cracking*. Straightforward polymerization yields polyethylene, which is produced in two forms—low density, for food wrapping and the like, and high density, for bottles and other more robust applications. Perhaps less obvious is the product vinyl chloride, from which PVC is made. Other common materials sourced from ethylene include ethylene glycol, an antifreeze, and polyester resins for carpets and clothing. Styrene is a derivative, from which polystyrene resins are made for containers, and also butadiene elastomer for tires. Readily apparent is the fact that ethylene is the starting point for a number of commonplace products. Consequently, increased domestic production of this chemical can have a major impact on much of the chemical industry. High-value exports could be the result.

Prior to this windfall of ethane availability, ethylene was largely produced by cracking naphtha, a byproduct of oil refining. These crackers have the ability to make gasoline or ethylene, with the proportions driven by relative pricing. When ethylene is made from naphtha, a significant fraction of the output is propylene, another important building block for the chemical industry. When ethylene is made from ethane, this fraction is not available. So, abundant ethane and subsequent cracking to ethylene will create a shortfall in propylene. The consequences of this are discussed in chapter 13.

Ethane Pricing

Ethane pricing has demonstrated an unusual pattern since 2009, entirely mediated by the shale gas boom. Prior to that year ethane tracked with oil price. This can be seen in Figure 8. The NGL composite includes ethane, propane, butane, and natural gasoline. It, too, tracked oil and generally stuck close to the WTI price. But from 2009 onward, much wet gas was produced, and ethane made up nearly half of the NGL produced from wet gas. This sudden increase in supply without a concomitant increase in demand caused the ethane price to drop while the larger molecules such as propane and butane stayed high.

As is evident in Figure 8, this separation became marked from the start of 2010. This trend continued in 2011, and then in 2012 ethane took a major dive, as shown in Figure 9. Early in 2013 ethane was priced about the same as natural gas, which is completely unprecedented.

Figure 8. Comparative spot price movements in natural gas, crude oil, and natural gas liquids, 2008 to 2011

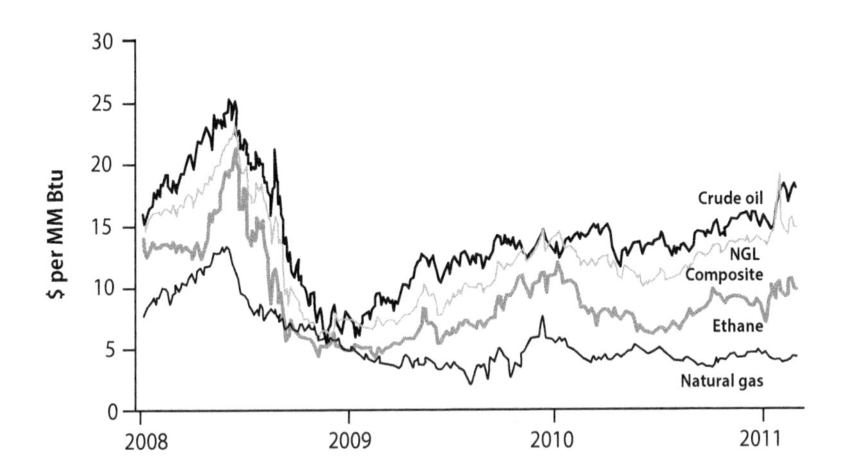

NGL composite includes ethane, propane, butane, and natural gasoline.
Source: US Energy Information Administration, April 29, 2011

Figure 9. Ethane price trend compared to oil price (WTI), 2008 to 2013

WTI refers to West Texas Intermediate, a benchmark for crude oil pricing.
Source: Figure courtesy of Dow Chemical Company, 2015.

Industry has responded to this trend by expanding cracker capacity and constructing new crackers. Range Resources in Pennsylvania is reported to have signed a contract to export ethane to Europe. This is very unusual because this gas is not very easily transported. This and all other demand creation will need a few years to take hold. Consequently, there is a school of thought, shared by me, that ethane prices will stay unusually low for four to five years and then drift upward, back to the pattern seen in the later years of Figure 8. A return to the prices of 2009 is very unlikely if natural gas prices stay under $5.50 per MM Btu for the next decade, as currently predicted by the EIA. With low prices, wet gas will continue to be the primary target of production, thus keeping ethane in abundance.

Impact of Low Ethane Price

Given a choice between naphtha and ethane as starting feedstock, the latter will be picked every time in North America because of the cost differential. The profit margin from producing ethylene from low-cost ethane is hundreds of dollars per tonne higher than from the use of naphtha as feed. As a percentage of ethylene prices in 2014 in the vicinity of $1,320 per tonne, a typical $600 differential with naphtha-sourced ethylene is highly material. Since cheap ethane is purely a North American phenomenon for now, the US is the low-cost ethylene producer in the world today.

Consistently low pricing combined with strong availability is changing the landscape of the North American chemical industry. In a 2011 report entitled "Shale Gas and New Petrochemicals Investment: Benefits for the Economy, Jobs, and US Manufacturing," the American Chemistry Council, a trade-supported organization, made some bold predictions. They predicted that ethane production would increase by 25 percent. They estimated a $32.8 billion increase in chemical production as a result, and an associated $132.4 billion increase in US economic output. Total estimated jobs created would be 395,000. In 2014 the same organization released more general statistics:

> As of [February 2014], 148 projects valued at $100.2 billion have been announced. These projects—new factories, expansions, and process changes to increase capacity—could lead to $81 billion per year in new chemical industry output and 637,000 permanent new jobs by 2023. More than half of the investment is by firms based outside the United States (American Chemistry Council, 2014).

The US now is extremely well positioned competitively with the rest of the world. Some evidence for this comes in the form of the statistic mentioned by the American Chemistry Council above: over half of the capital investment is from abroad. The unit cost of production of plastic resins in the US is now on par with Canada and only slightly higher than in the Middle East. It is well below all the other consuming nations, including China, Western Europe, and Japan, in ascending order of cost. In 2005 the US was on par with Western Europe. Significant investment that had fled to what had been low-cost countries is returning, contributing to the economic figures estimated by the American Chemistry Council. Imports will increasingly be replaced by exports.

Since all of this is so new, policymakers must feel they are in something of a fog. LNG producers such as Cheniere have obtained permission for exports, when not so long ago they were building import terminals. In early January 2012, Congressman Ed Markey of Massachusetts expressed concern to Energy Secretary Steven Chu that approval of the Cheniere request would result in increased domestic gas prices. That particular concern may not have had a lot of merit because of the sheer abundance and accessibility of the resource. But this does raise the question as to what is the most advantageous export: a commodity such as natural gas or a processed product? There can be little doubt that it ought to be the latter, if feasible. In chapters 16 and 19, I discuss the technical and economic viability of methane conversion to transport fuel. Ethane-derived resinous material, methane-originated fertilizer, and methane-sourced liquid fuel are the type of high-value products most suited to export. The bulk of the economic value stays in this country.

The Ethane Cracking Options

North America, excluding Mexico, has 42 ethane crackers. Of these, 33 are on the Gulf Coast, 4 in Alberta, 2 in Ontario, 2 in the Midwest, and 1 small one in Kentucky. So, the Bakken Shale oil field could supply Alberta, while Eagle Ford in south Texas, a rich source of ethane, is proximal to Gulf Coast capacity. But the Marcellus and Utica are essentially stranded. These are also generally acknowledged as the largest of the deposits. This analysis concentrates on these two sources.

The various options are presented in Figure 10. I will key in on only the first, third, and fourth rows from the top. The most conventional option is to transport the wet gas to a central facility and chill it to the point where all the NGLs drop out sequentially. The ethane is the last to liquefy and can be separated and then transported via pipeline to a cracker. As mentioned earlier, the only viable capacity is on the Gulf of Mexico coast. Much of this capacity was idled in the early part of this century following the gas price shocks and fled, primarily to the Middle East.

Figure 10. Processing technology options for shale gas ethane

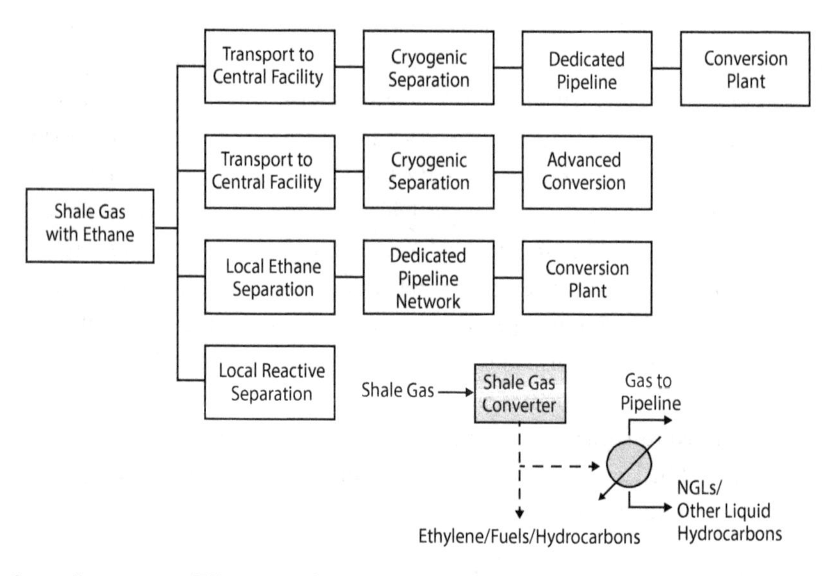

Source: Figure courtesy RTI International.

Another option is to separate the ethane at or near the production location and then pipe it to a central cracker in the region. So, for example, a dedicated cracker in eastern Ohio or West Virginia or western Pennsylvania could serve all of those areas.

The final option is the most innovative and also provides the highest returns. I refer to it as the *pad-level (local) reactive separation option.* This option involves reacting all the gases at the wellhead and tailoring the reaction to crack or convert the ethane (ethylene/fuels/hydrocarbons in Figure 10) while allowing recovery and sale of the other liquids. The methane passes through to the export pipeline.

The Gulf Coast Cracking Option

In late 2011, El Paso Energy dropped plans for an accelerated-schedule pipeline that included using an existing portion. This was because the company was unable to get guaranteed contracts for gas. All major capital undertakings such as pipelines and LNG plants suffer from the need for these assurances of volume usage. But at about the same time, Chesapeake Energy, a major producer of gas, announced its decision to ship 75,000 barrels per day (ultimate capacity of 120,000) of ethane to Mont Belvieu, Texas, a distance of about 1,230 miles. West Virginia authorities responded with dismay.

"Every barrel of ethane that is shipped to the Gulf Coast means jobs and investment there and not here," West Virginia Department of Commerce Secretary Keith Burdette said. "We want to see those investments in this part of the country." But he acknowledged that absent an in-state option, producers have no option but to build a pipeline to a known market. The downstream oil and gas establishments clearly wish for a Gulf Coast cracker option. State governments will equally vehemently opt for the local option. In the end, the timeliest solution may be the winner. This may be the opportunity for the pad-level (local) reactive separation option, provided it can be developed in timely fashion.

More recently, MarkWest and Kinder Morgan announced plans in early 2013 to open an 1,100-mile NGL pipeline from the Utica shale well fields to the Gulf Coast. They also plan smaller lines to bring the NGLs to a location to process the mixture to strip the ethane and place it in the line to the Gulf. Recently completed in Ohio is the Appalachia-to-Texas (ATEX) Express Pipeline built by Houston-based Enterprise Products Partners L.P. to transport liquids from the Utica south. When others come on stream in addition to this one, the ethane pricing ought to normalize closer to the numbers in the vicinity of 2010 in Figure 9.

Overall capital costs for such conventional ethane recovery, transportation, and conversion would be on the order of $2,700 per annual tonne of ethane (assuming construction of a new 1 million-tonne-per-year ethylene cracker on the US Gulf Coast). Overall annual operating costs for such conventional ethane recovery, transportation, and conversion would be on the order of $190 per tonne of ethane. Total annual costs of conventional ethylene production via shale gas extraction, ethane recovery, ethane pipelining to the US Gulf Coast, and ethane cracking would be expected to be on the order of $1,100 per tonne of ethylene (including capital recovery and shale gas extraction costs).

This is in the right ballpark for current ethylene pricing, indicating such a conventional approach based on shale gas could be competitive with oil-based ethylene.

The New Proximal Cracker Option

This option is seriously being considered by Shell, among others. However, in late 2013 Shell announced a seemingly indefinite postponement. West Virginia authorities are keen to realize this option, and Pennsylvania and Ohio would like this as well, in their states of course. This option could reduce overall pipeline capital and operating costs by as much as 90 percent versus transporting ethane to the US Gulf Coast. Overall capital costs for this option would be on the order of $1,600 per annual tonne of ethane. Overall annual operating costs for this option would be on the order of $130 per tonne of ethane. This is decidedly cheaper than the Gulf Coast option, although the figure for the latter could be shaved with use of some existing facilities. More to the point, this option would require less pipeline cost and associated community pushbacks similar to the issues with the Keystone XL heavy oil pipeline, which is still not resolved in 2014.

But the most important reason for this option is job creation in general proximity to the production. Not to be ignored also is the fact that the highest density of polyethylene converters to useful products is in the Midwest. The Gulf Coast option would entail shipping the polyethylene back to near where the ethane came from. Most modern crackers are fully integrated to produce all the necessary downstream plastics. One could reasonably expect the proximal cracker to serve the polyethylene and other plastic needs of the East and Midwest. This option allows all the job and value creation to be proximal to gas production.

The Pad-Level Reactive Separation Conversion Option

A proposal under consideration is reactive separation conversion of ethane at the well pad level. It would need to have the following features. Based upon the assumptions shown in chapter 13 for Marcellus prospects, such a converter would need to be economic for throughputs of between 2,000 and 3,000 barrels of ethane processed per day. The output at a minimum must be ethylene, but more preferably a drop-in fuel for gasoline. That is a tall order for any conventional cracking process, but new technology appears available to achieve this, and RTI International believes it is feasible (R. Gupta, personal

communication, March 13, 2012). Such a small-footprint convertor could be expected to be easier to finance and commission than a full-blown cracker.

The full size would be on par with the pilot plant for a regular cracker, so time to market can be expected to be shorter than for most chemical processes. Timeliness with regard to jobs and the thwarting of raw ethane export ought to make it a favorite of state authorities.

The prominent producers in the Marcellus and Utica are not players in the downstream refining and processing area, so their imperative ought to be prompt monetization of the ethane. Currently, producing states are unhappy about plans to ship ethane to the Gulf but reluctant to antagonize a major employer. But given a viable local option, they may exert pressure and possibly provide investment incentives. To date the only local option available to them is a new dedicated cracker, and that may be in a neighboring state; also, the size and capital cost inevitably delay things.

The biggest hurdle, assuming techno-economic viability, is the business model. A large integrated cracker producing polyethylene is the current model. The independent polyethylene manufacturer is an endangered species. The reactive separation approach will require polyethylene to be manufactured in reasonable proximity. This is because the polymer is much cheaper to transport than is ethylene. The combined economics of all processes culminating in polyethylene will be the key, in comparison with the local cracker option. Of course, if the technical hurdle of the economic production of a drop-in fuel locally is crossed, that could trump all other options.

The solution to the ethane dilemma will most likely hinge on a combination of innovation and state activism.

CHAPTER 15

The Alaska Pipeline Is Dead; Long Live the Alaska Pipeline

"The leader of the band is tired / And his eyes are growing old"

—From "Leader of the Band" by Dan Fogelberg

Alaskan oil represented 11 percent of US consumption in 2012 and is a bit less today due to shale oil, but it remains a vital component considering that the US imported over 50 percent of its need in 2011, even though that proportion is dropping. The Trans-Alaska Pipeline System (TAPS), a marvel of engineering built in 1977, is now in trouble. Another proposed pipeline, one to transport natural gas to the Lower 48, is on life support. This is the story of one that needs resuscitation and one that needs to be allowed to die. Curiously, these two outcomes are related.

A giant oil field was discovered in 1968 in Alaska. But it was located up on the frozen North Slope. The sheer size and national priority dictated the construction of a pipeline (Figure 11). This was no mean task. The 800-mile traverse included three mountain ranges and hundreds of rivers and streams. Over half the length needed to be well above ground in order to not melt the permafrost, ground that is permanently frozen. The melting would occur because the oil flowing is relatively hot and needs to remain so in order to flow at a good rate. The above-ground pipe also needed to be built taking into consideration caribou migration and the like.

TAPS is in trouble. At its peak it carried 2 million barrels of oil a day (MM bpd), and the oil made the trip in three days, arriving at a temperature of about 100°F. Now, the falloff in production has reduced the flow to a third of that and it takes up to 15 days of travel. Figure 12 shows historical flows of crude oil in TAPS plus two projections: a continuous decline and a case where measures such as pipeline heating improve the flow.

Figure 11. Map of Trans-Alaska Pipeline System (TAPS)

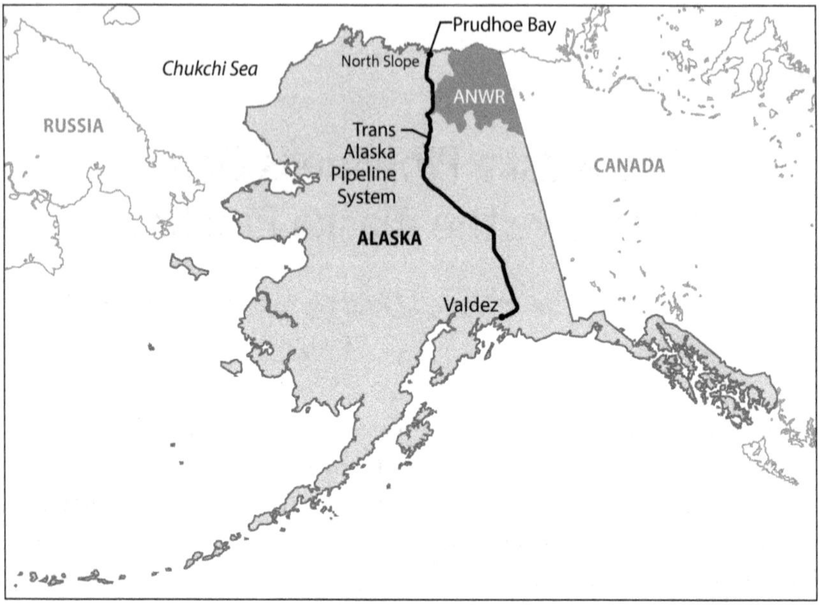

Prudhoe Bay
Chukchi Sea
North Slope
ANWR
RUSSIA
Trans Alaska Pipeline System
CANADA
ALASKA
Valdez

Source: Figure courtesy RTI International

Figure 12. Average daily volume of the Trans-Alaska Pipeline System, in thousands of barrels per day

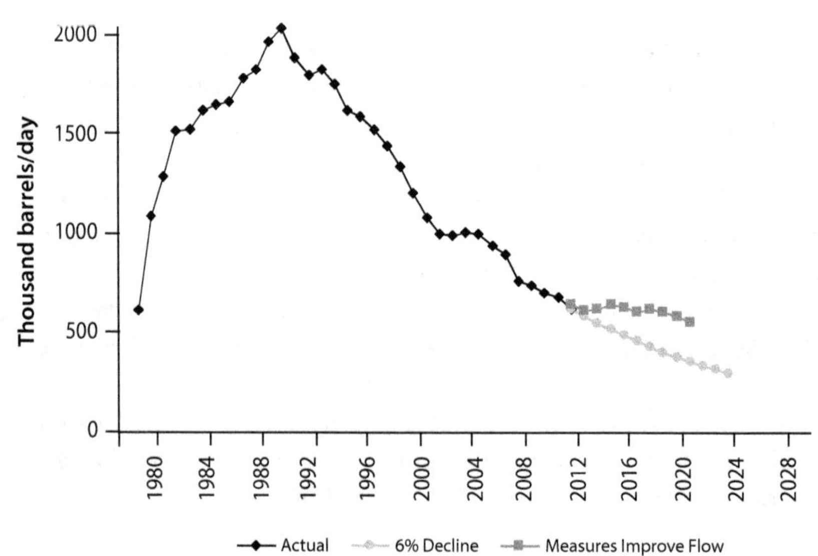

Source: Alyeska Pipeline Service Co., 2010.

The heat loss with the extended travel time causes the oil to be much cooler upon arrival. This increases the possibility of clogging and other problems. The practical limit at which it becomes economically prohibitive is debated but is somewhere in the range of 0.35 to 0.5 MM bpd, although the lower end of that range would need heat management solutions. That day is a scant few years away, 10 at most.

More oil is needed. The three candidates are heavy oil, oil from the source rock, and liquids converted from natural gas (called gas-to-liquid, GTL). Offshore oil from the Chukchi Sea and oil from ANWR (the Arctic National Wildlife Refuge) will not be considered here because politics, environmental hurdles, and location will render the output not timely for resolution of this problem.

Heavy Oil

The Ugnu formation is estimated to hold about 20 billion barrels. The reservoirs are very close to the permafrost layer, so the oil is cold and hard to move. BP's Milne Point development has already demonstrated the ability to produce economically viable quantities using cold production methods. Such methods rarely yield more than 8 to 10 percent of the fluid in place. But even with this low recovery, the production will at least serve the purpose discussed above. Eventually BP will try some heating method because the layers adjacent the permafrost will be prohibitively cold. These methods necessarily will need to avoid damaging the permafrost. This is likely feasible.

The stuff is the consistency of molasses and not easy to transport. It would need blending with a lighter oil of some sort to be suitable for TAPS. The current oil from the Slope would do, as would condensate from the gas production. Another heavy oil source is the West Sak in the Kuparuk River field, also on the Slope. This has like quantities and is a bit lighter, more the consistency of honey. It too is close to existing infrastructure.

Heavy oil has a higher carbon-to-hydrogen ratio than does light oil. When refined, a significant portion of heavy oil drops out as a carbon residue known as petroleum coke, which has low value. Consequently, heavy oil sells at a discount. Heavy oils from Canada and Venezuela, major import sources to the US, also suffer from high sulfur and heavy metals. Alaska oil is much better in this regard. Nevertheless, only specialized refineries can take heavy oil. This will eventually prove to be a hurdle when the volumes down the TAPS are high. The good news is that California refineries, the closest to the Alaska port of Valdez, are generally well positioned to handle heavy oil refining.

Oil From Source Rock

The concept here is similar to our discussion with respect to shale gas. Proponents suggest that only about 20 percent of the oil "at source" found a suitable rock into which to migrate. They maintain that the remainder can be tapped much as is shale gas today, by fracturing the rock. In my discussion of the origins of shale gas, I, too, conjectured that more would stay behind than would find a porous and permeable home. Now a little outfit out of nowhere, Great Bear Petroleum, has spent $8 million on substantial acreage. They will first target the Shublik formation, rock that is about 250 million years old and is the deepest of the three source rocks in the area. In part this target must be chosen because the deeper rock is more naturally pressured. Great Bear expects to access the shallower deposits in a later phase. This is not unlike my discussion with regard to developing the Utica first and then the overlying Marcellus.

The president of Great Bear has a BP (actually Sohio, back when he was there) Alaska pedigree as a geologist. So, this lends him some credibility. Great Bear had expected to start production in 2013 and ramp up to 250,000 bpd by 2020. That would be ample to save TAPS. However, as of mid-2014 they had just two exploratory wells and hadn't yet taken either horizontal. They have spent $40 million on seismic exploration and appear encouraged by the results. Two other companies have purchased leases close by, but all eyes are on Great Bear.

There is every reason to believe that the source has the same mix as Eagle Ford and Utica: zones of oil, wet gas, and dry gas. One assumes Great Bear is chasing the oil leg first. But either their acreage or those adjacent must have the other two, which will have bearing on the following discussion. It is interesting that none of the current North Slope producers bid for the leases (except for a solitary lease by ConocoPhillips). The post–shale gas world knows all about source rock recovery. So, one school of thought would be that they will wait for success before bidding on adjacent leases yet to come up for sale. Or they plan to simply buy Great Bear, following a Lower 48 pattern in shale gas.

Finally, there is the ghost of Mukluk, which was predicted to be a giant reservoir. The company: Sohio. When the drillers got there, a billion (1983) dollars later, the oil appeared to have been spirited away. Dry as a fossilized bone. There is no geological parallel to the matter above. Just the memory of the last big one that got away. And this one is also big.

Shale Gas Kills the Gas Pipeline

There always has been natural gas in association with the oil in Alaska. With no ready market, it is pumped back in the ground to the tune of 8 billion cubic feet (bcf) per day. Estimates of gas reserves vary, but there are at least 35 trillion cubic feet in the vicinity of Prudhoe Bay. Another 100 trillion are estimated as potential for all the other areas, including the Beaufort and Chukchi seas and ANWR. For conversion to oil equivalent, one roughly divides by 6,000. So the total likely is about 22 billion barrels of oil equivalent. That is close to the light oil estimate for Alaska. The point is that the gas is a world-class resource and presently without a market.

The perceived solution has been a gas pipeline to the Lower 48. At least it was until shale gas came along. All of a sudden cheap shale gas rendered the Alaskan gas completely unnecessary. Fortunately for everybody, the bickering delayed the decision to this point when it is truly passé. According to the National Energy Technology Laboratory (National Energy Technology Laboratory [NETL], 2009), a gas pipeline would require daily throughput of 4.5 billion cubic feet per day for 35 years. That translates into a reserve of about 59 trillion cubic feet, which you will note is more than the readily accessible 35 trillion mentioned earlier. But in the end that is hardly the point. The Lower 48 is expected to be self-sufficient, and there is already concern regarding prices staying too low. Alaskan gas would not help that. But there is a remedy that leaves us with a TAPS that is saved rather than played on a trumpet solo.

Liquefied natural gas (LNG) export is a new element for Alaska. The ConocoPhillips LNG plant in the Cook Inlet was granted an extension in 2013. In early, 2014 the Alaska governor announced completion of a study to send 3.5 bcf per day down to the Cook Inlet with intent to export LNG. I believe much of the course would follow the existing oil pipeline. That volume would feed two substantial LNG plants, much bigger than the existing one. This is an interesting development and with a lot of merit. This is the only LNG export idea that has zero impact on the Henry Hub gas price. This gas is essentially stranded in Alaska.

The Gas-to-Liquids Option

A full discussion of converting gas to liquids in the context of cheap gas is in the following chapter, "Transport Liquids from Gas: Economical Now." But in short, it involves first reacting the methane with oxygen or steam to produce a synthesis gas, a combination of carbon monoxide and hydrogen.

This reaction is straightforward, and the reaction products can also be obtained with coal as a starting point. The tricky part is then catalytically converting this mixture into long-chain hydrocarbons, also as described in chapter 16. During World War II, virtually the entire war effort was run on transport fuel using the Fischer-Tropsch (F-T) process. The Germans had domestic coal but not oil. Closer to the present day is the optimization of this process by the South Africans. During the apartheid-driven embargo, South African oil companies were forced to develop fuel with domestic sources. Today Sasol is arguably the leading purveyor of this technology, although the likes of Shell would certainly argue that point.

ExxonMobil, one of the big Prudhoe Bay property owners, has a process for taking synthesis gas to methanol and thence to gasoline, the so-called MTG process. But whether F-T is used or MTG, the prospect of a liquid fuel from Alaskan gas is real. The gas itself ought to be priced very low. This is particularly the case if it is gas associated with oil production. Using it means not incurring the cost of putting it back in the ground. One could even argue, tongue planted close to the cheek, that the price is negative.

The key question is how much gas capacity is needed to support a commercial-scale GTL plant. A plant producing 100,000 bpd would be materially useful to keep TAPS open. The gas required for this is about 1 billion cubic feet per day. That is a far easier target than the 4.5 billion required for a gas pipeline, even were it to make economic sense, which it does not due to shale gas in the Lower 48. A comfortable target would be a 200,000-bpd GTL facility. Note also that the industry currently reinjects 8 bcf per day.

Things start to get interesting if one combines 200,000 bpd GTL output with a like quantity of heavy oil. In some ways this is near sacrilegious! GTL-produced liquids are incredibly clean; they are free of sulfur and all manner of impurities present in the same liquids from oil refining. But, getting past this squeamishness, the light fractions from a GTL plant would be very well suited to blending with heavy oil to make the latter transportable.

There are issues one would have to deal with. Sometimes when heavy oil is mixed with certain light hydrocarbon fractions, some of the carbon in the heavy oil will precipitate out as asphaltenes. You don't want that happening in the pipeline. But this sort of thing is well understood and simply needs to be handled. In fact, one option would be to use light molecules associated with the natural gas to deliberately precipitate the asphaltenes prior to shipping. A commercial process exists for doing this and is known as the ROSE process. In that case the GTL-derived product would simply be sent down the pipeline in batches between other oil batches.

The Trans-Alaska Pipeline System is critical for US energy security. It must be kept open. There is currently a lot of rhetoric to the effect that this specter is being used by oil interests to open up environmentally sensitive areas. A future that combines production of heavy oil, blended with a liquid from natural gas, is one that we can all live with. This solution ought to keep the pipeline open for decades. If the shale oil option is realized, that would just be icing on the proverbial cake. That would merely raise the flow in TAPS back to the glory days' numbers. Alaska gas pipeline: rest in peace. A lone bugler can play "Taps."

CHAPTER 16

Transport Liquids from Gas: Economical Now

"I can't get no satisfaction / 'Cause I try and I try and I try . . ."

—From "(I Can't Get No) Satisfaction" by The Rolling Stones (written by Mick Jagger and Keith Richards)

Economical conversion of gas to liquids (GTL) has been repeatedly attempted by every major oil company over the years and always come up short of expectations. Shale gas now offers up a future with more certainty on gas pricing in North America, and possibly other parts of the world as well. Until now the only viable sources of gas for GTL were Qatar and Iran. The vast resource base, predictably low delivered cost, and distance from consumer gas markets made conversion of gas to liquids relatively attractive. All three of these conditions apply to Alaskan gas and the first two to shale gas in the Lower 48.

Expectations were high with the efforts of Shell and Sasol separately with cheap Qatari gas, but capital cost overruns dogged those projects, especially for Shell. The timing was unfortunate in that the majority of the construction was contracted during the prerecession boom and associated higher costs. There were also some reported teething problems on the process side. One assumes that at least these two companies have a handle on doing it more cost effectively the next time.

Industry kept pecking away with innovative catalytic steps, and today cheap shale gas may enable economical production in North America. Certainly abundant low-cost stranded gas in Alaska is fair game for the conversion, as discussed in the previous chapter.

In the late 1920s two Germans, Franz Fischer and Hans Tropsch, invented the process named after them and commonly abbreviated to F-T. A singular lack of oil in Germany combined with abundant coal led them to devise a means to obtain transport fuel from coal. This coal-to-liquids (CTL) process continued to be refined, but primarily for the gas-to-liquids (GTL) embodiment. The exception was Sasol in South Africa, which continued on the coal-derived path in response to the apartheid-induced embargo in 1987, which prevented oil imports. The Sasol work began well before that date in part due to the scarcity induced by the Arab oil embargo in 1973. As shown in the box, after the initial step of production of synthesis gas or syngas, the process steps are exactly the same for CTL and GTL.

Basic Principles of the Fischer-Tropsch Process

The starting point for these reactions is some carbonaceous material and a source of hydrogen. In the original discovery, coal was reacted with water in the following manner: $H_2O + C \rightarrow H_2 + CO$. This is sometimes referred to as the water gas reaction. The reaction product is designated synthesis gas, or syngas for short. Syngas is a basic building block for a number of final uses, including simply combusting to produce electricity. In that case, often the reaction is taken further, as follows: $H_2O + CO \rightarrow H_2 + CO_2$. The hydrogen is combusted for power, and the CO_2 is disposed of in some way. If the CO_2 is sequestered, this process is the basis for clean electricity from coal, as in the federally funded FutureGen program.

The syngas may also be produced by oxidation of methane according to the following reaction: $H_2O + CH_4 \rightarrow CO + 3H_2$. This process is known as steam reforming.

In the F-T process, the syngas produced is tailored to a ratio of H_2:CO depending on the catalyst used, usually close to 2:1. It is then reacted according to the following equation:

$(2n+1)H_2 + nCO \rightarrow CnH_{(2n+2)} + nH_2O$, where n is an integer.

At $n = 1$, we get methane, which is certainly not desirable because that was our starting point in the case of natural gas as the feedstock. The key chemical challenge is creating the carbon-to-carbon bond, and this is more challenging for higher *n*'s. The higher the *n*, the more energy-dense the liquid. Gasoline has *n* about 8, diesel around 12, and jet fuel in the range 10–15. All these fuels are mixtures with *n* as a range rather than a single number. The conditions of temperature and pressure determine the output. This reaction is very exothermic, that is, it generates a lot of heat, and this is a key process control variable.

F-T is the most versatile of the different methods for converting methane to liquids. This is because process conditions can be adjusted depending on the output desired. The starting feed for the gasification to syngas can be biomass as well, although typically preparation of the feed to be suitable for a gasifier would be a precursor step. Although not the subject of this essay, a new technology for converting biomass to transport fuel is worth noting. This is known as catalytic pyrolysis and involves treatment to a state short of actual combustion, resulting in the formation of a liquid. This liquid is very similar to crude oil and could be a direct feed to an oil refinery.

The MTG Process

One of the other technologies is the so-called methanol-to-gasoline (MTG) process conceived by Mobil about 40 years ago. This involves first going from syngas to methanol, which is a very simple process, and thence to dimethyl ether and finally to gasoline, which is a more complex catalytic process utilizing a zeolite designated ZSM-5. This reaction, too, is highly exothermic, so efficient removal of heat is a key.

The output is a gasoline ready to mix with conventionally produced gasoline. The MTG process was commercially operated in New Zealand by Methanex New Zealand Ltd. from about 1979 to 1996. The output was a gasoline with octane rating in the low 90s, in other words like the highest grades at our pumps today. The plant was then mothballed in 1996, with economics given as the reason. Since then ExxonMobil has improved the process and is offering it commercially today. The denser fuels (diesel and jet fuel) cannot be synthesized using this process. However, since gasoline is such a large part of the passenger car fleet, this is not an undue limitation.

Economics of Conversion

Consider a plant producing 65,000 bpd of output. For the statistics here I use the analysis of Michael Economides (Economides, 2005) for the input/output figures:

Input per day:

Natural gas:	685 million cubic feet

Output per day:

Diesel:	44,000 barrels
Naphtha:	17,000 barrels
LPG:	4,000 barrels

The selling price of the components can be highly variable. All refined fuels such as diesel have a value over that of crude oil which is known as the *crack spread*. The term originates from the fact that large crude oil molecules are "cracked" to the smaller molecules that make up diesel or gasoline. The term *n* in the formula shown in the box is greater than about 22 for crude oil, and for gasoline it is around 8. The cost of this cracking plus a profit is termed the *spread.* This varies a lot over time and is one of the reasons that the price of gasoline at the pump sometimes appears not to be affected by the price changes of crude oil.

For diesel, a representative figure for the crack spread would be $15 to $20 per barrel. The spread for naphtha I'll estimate at about $5, and LPG I'll assume to be the same as light sweet crude, in other words zero spread. Based on these assumptions, and a crude oil price of $100, the value of the output is $7.24 million. The cost of the gas, at a price of $4 per mcf, is $2.74 million. Assuming a cost of $20 per barrel of output as the total cost of depreciation and operations, that adds $1.3 million to the cost, bringing the total cost to about $4.04 million, leaving a total profit of about $3.20 million. This is a very healthy margin, even assuming costs are higher in some places. But the main point is that it is sensitive to gas price. If the price goes to $12, as it did on occasion in the last decade, the cost jumps to $9.52, which is a severe loss position. Even $8 makes it marginal as a business.

This underlines the fact that shale gas–enabled consistently low prices would be a key to the business decision to build conventional GTL plants in North America. The other variable in play is the cost of light sweet crude. I assumed $100 in the analysis. Elsewhere in the book I support the models that predict oil prices as being consistently high and gas prices low. This is a recipe for widespread GTL. Certainly Alaska, with essentially stranded gas, is a prime candidate. Construction costs will be higher there, but the gas prices can be assumed to be a good deal lower than prices in the Lower 48.

Analyses by the National Energy Technology Laboratory have shown that liquids synthesized in Alaska could be sent down the existing oil pipeline in slugs between crude flows. I also suggest in chapter 15 that the transport of heavy viscous crude could be enabled by appropriate blending with GTL liquid.

Drop-In Fuels

This is a term that has come into vogue to indicate a synthesized fuel that can blend in with gasoline or diesel, or in some cases replace it. In some ways ethanol fits the description in that it blends with gasoline, but it has less energy density and so impairs a car's driving range. Butanol, on the other hand, would come close to being a true drop-in. A number of efforts are under way to produce such fuel from sugar ("The Future of Biofuels," 2010). I'll restrict my discussion here to fuels using methane as the feed.

F-T plants are required to be large to achieve economies of scale. The cost runs into billions. The Shell facility in Qatar produces 260,000 barrels of total fluid output per day and cost $18 billion to build. Future plants could be somewhat lower in cost but not dramatically so. This has two implications. One is that a large volume of gas has to be contracted for at least 20 years, and there has to be reasonable assurance of cost stability. The other is that the sheer size of the investment makes financing difficult. Consequently, industrial research has continually been searching for that elusive small-scale process that is cost-effective.

Oxidative Coupling of Methane

An approach that has been researched considerably is to attempt to directly couple two molecules of methane to produce a carbon-to-carbon bond of a larger molecule, ethylene. The desired reaction is:

$$2CH_4 \rightarrow C_2H_4 + 2H_2$$

This is highly exothermic, and temperature control is a key. The ethylene can be used for a number of applications, but further steps need to be taken to convert it to longer-chain molecules that can drop into gasoline. This is the goal of GTL.

This reaction has a low yield, and some component of ethane will also be produced. But the low yield is what has held back this approach. Now one company, Siluria, claims to have developed a unique catalyst to dramatically improve yields. Its website describes a process for creating a catalyst using an interesting amalgam of organic biochemistry and nano-catalyst dispersion in such a way as to yield a structure with highly reactive surfaces. If this indeed works as claimed, it will put oxidative coupling of methane into the commercial arena. One would then expect that modestly long-chain drop-in molecules could be produced for gasoline using much less energy than is needed for conventional GTL methods.

Conversion to ethylene and then to alkanes is one approach. Others are using advanced catalysis to go directly to diesel lookalikes. Yet others are working on controlling the kinetics using proven catalysts. The descriptor "game changer" is used with much abandon in the popular press, right up there with "paradigm shift." Here, though, is the prospect of a technology that could truly change the game. The promise here is of low-cost conversion with a small footprint. This last is the key. Even if the cost were to be the same as conventional GTL, the ability to conduct distributed processing at small scales is what is seductive.

In early 2014 at least a dozen companies were pursuing this objective. In a recent World Bank report (Fleisch, 2014), Theo Fleisch reviewed these, with the imperative being amelioration of the flared gas problem. He believes that four companies are close to commercialization: Velocys, CompactGTL, and Greyrock, all F-T targeting diesel, and Oberon Fuels, producing DME. All the diesel producers aim to have an output with a minimum of long molecules known as wax. This is because hydrocracking wax back to diesel is expensive. Also, not having that back-end process reduces the complexity of the plant. This is especially important for processes in field locations.

This objective of limiting molecule size is best explained by Figure 13, which is an output from a research program at RTI International. Shown in Figure 13 is the weight percent distribution of molecules in the reactor output plotted against *n*, in the formula C_nH_{2n+2}, which is the basic formula for all oil constituents, as explained in chapter 1. The diamond points represent the distribution of molecules in the standard GTL process. One can see that a significant portion of the output is in the wax range of *n* greater than about 25. The square points are from an experimental process attempting to minimize the production of wax. Diesel, the desired fuel output of the process, is a mixture with *n* ranging from about 12 to about 18. As noted, a very small

Figure 13. Distribution of molecule size in a process aiming to minimize wax production in the Fischer-Tropsch process

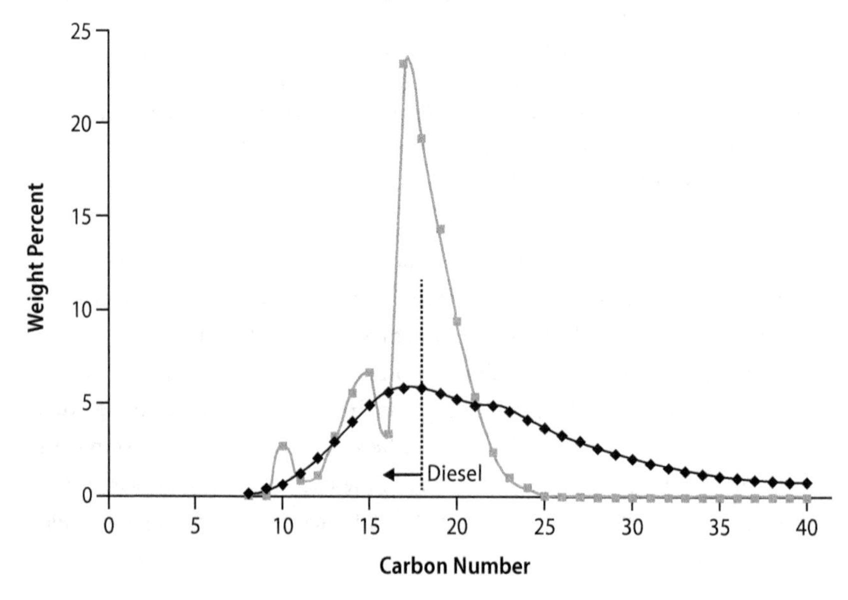

Figure courtesy of RTI International.

fraction of the output has molecules in the wax region. All four companies mentioned by the World Bank report and cited above have this objective in mind.

The ability to economically produce transport liquids from natural gas on a distributed basis has two impacts. One is the augmentation of conventional sources with transport fuel produced closer to the consumption, with minimal transport of the gas. The explosion of gas production has strained the pipeline systems, and utilizing at least a part of the gas locally will ease the problem. This will be particularly apropos in countries such as Poland and India, which don't have extensive pipelines. The second advantage is that the small scale of each plant eases financing and does not require massive gas supply contracts to be in place. As a bonus, these plants will be in gas-producing states, keeping the economic value local.

CHAPTER 17

Natural Gas Vehicles: A Step in the Right Direction

"All in all it's just another brick in the wall"

—From "Another Brick in the Wall" by Pink Floyd (written by Roger Waters)

Direct use of natural gas as a fuel for vehicles could have a material impact on the reduction of imported oil. This is technically feasible and likely most applicable to commercial and fleet vehicles, especially for long-haul trucks. Light-duty passenger vehicle applications will need innovation to overcome some consumer-unfriendly features.

Some see natural gas as a transitional fuel toward a low-carbon future for North America. This can be manifest either simply in displacing coal for electricity production or in displacing imported oil for transportation purposes. The former reduces carbon dioxide emissions by virtue of the fact that the combustion of natural gas produces about half as much greenhouse gas as coal does. There is little argument on this point, except some argue that the act of producing the gas creates fugitive methane with adverse consequences; that is discussed in chapter 9. The more interesting avenue, substituting for oil in transport applications, creates debate. One approach would be to simply use the cleaner electricity to drive electric motors in cars.

Comparison to Electric Vehicles

The well/mine-to-wheel efficiency of an electric car is far in excess of any internal combustion engine, no matter what the fuel. So, simply in terms of using the least energy to drive a given number of miles, an electric car is the best option (see box on the following page for the calculation). This addresses a point that is not made often enough: we need to figure out how to get the same amount of gratification with less energy. Then where the energy comes from is less of an issue because associated emissions will be reduced, as a result of simply using less energy.

For example, today, due in large part to a California initiative commenced in 1971, television sets have standby power use of less than 1 watt. This is the power to simply keep the device in ready mode to allow the use of the remote monitor without a hard on/off at the machine. This little bit of couch potato convenience used to cost up to 12 watts. Similar standby wastage is in evidence in power drawn by devices left plugged in ready for use. This includes cell

Electric Cars Use Less Energy

To give support to this assertion, I will calculate here the efficiency of each step in the process and arrive at a fair comparison. I'll use the following facts and assumptions:

- A gallon of gasoline has 116,100 Btu, which equals 34 kWh.
- The average car being replaced delivers 35 miles per gallon (I am being generous here).
- For years the dogma has been that electric vehicles (EVs) use 0.2 kWh per mile. Nissan reports that the Leaf averages 0.25 kWh per mile. As in all electric and hybrid cars, stop-and-go driving gives better mileage than continuous operation. So, that number could be higher in some cases. I will use the 0.25 number for this exercise.
- Refining oil to produce gasoline consumes 20 percent of the energy in the oil.
- Coal-fired plants have efficiency of 40 percent (60 percent energy loss); by using coal, not gas, I am being conservative, and this figure is that of newer supercritical combustors.
- Electricity lost in transmission is 8 percent (a good estimate for the US).
- Energy to get the oil out of the ground is a wash with coal mining. Had I used the less conservative gas source for electricity, the offset would have been precisely correct.

So, energy losses for gasoline prior to its being consumed in the vehicle are 20 percent. Energy used after combustion is: 34 kWh in a gallon divided by 35 miles to the gallon, further divided by 0.8, equals **1.25 kWh per mile.**

Energy losses for EVs are 60 percent at the generating plant, minus 8 percent in transmission, equals 32 percent. Energy used by EVs equals 0.25/0.32 equals **0.78 kWh per mile.**

The ratio of the energy used to drive an average gasoline engine car to that used to drive an EV is 1.25 to 0.78, or 1.6. In other words, a conventional vehicle uses 60 percent more energy as an EV for the same purpose. Is this exactly right? Probably not, but it is not off by much. The key takeaway remains that the EV advantage has a facet that is not commonly recognized in quantitative terms.

phone chargers—but the worst offenders are printers. Overall about 8 percent of US power usage is for this bit of convenience.

But a fleet turnover to electric vehicles will take decades. In the meantime it would be well to also have alternatives capitalizing on cheap and abundant natural gas. One is processing it to produce liquids that drop right in as replacements for gasoline or diesel. This is viable and is known as gas-to-liquids, or GTL, discussed in chapter 16.

A potentially important automotive fuel derivative of natural gas is methanol. The most effective use will likely be in the form of M85 (85 percent methanol, balance gasoline). Once again natural gas must be seen as a bridging raw material. Unlike ethanol, production from biomass is very straightforward. But if shale gas remains cheap it may be hard to displace as the primary source for methanol. In turn, gasoline substitution will be very economical.

This chapter is devoted to a discussion of the pros and cons of using natural gas directly in existing or modified internal combustion engines. But one message is clear from the calculation in the box: other things such as cost and range anxiety being nearly equal, an electric vehicle is by far the preferred option from the standpoint of emissions. Not only is it more efficient, as shown, but the tailpipe emissions are zero. Sure, the electricity producer emits carbon dioxide, but capture at a plant is more tractable than on each vehicle. Having said that, the latter is not completely infeasible, and one attempt at doing so is being researched.

Natural gas for cars and buses is generally in the form of compressed natural gas (CNG). The gas is compressed to a pressure of about 3,600 pounds per square inch. This is about 250 times atmospheric pressure. The tank required for this has to be robust, which adds weight but makes it safe on impact. Also, research and development is ongoing to minimize weight and cost. A promising avenue is the use of adsorbents to store the gas at volumetric densities of a factor of 2 over CNG. The energy density is also lower than that of gasoline by about a factor of 4. Combined with the weight penalty, this causes range to be reduced. The only car designed to run only on CNG, the Honda Civic GX, has a range of something over 200 miles, depending on how you drive. That range is double that of the all-electric Nissan Leaf.

A purpose-designed vehicle such as the Honda Civic GX could well have the tank below the trunk, although Honda did not do that for the current model. However, retrofit vehicles will need to take up trunk space, and the space occupied is significant. Yet this has been done in taxis in New Delhi and Kuala Lumpur. They simply install roof racks for luggage. The retrofit market is of interest for any quick uptake of this technology. The rate of uptake will determine the speed of installation of infrastructure.

Refuel Methods

Ease of refueling will be a key to acceptance. The uptake is much more likely to be swift for commercial vehicles than passenger. In many such cases, fleet refueling may be achieved simply on a dedicated basis. This has been the case for the cities in the world where all public transport has gone to CNG by fiat. Incidentally, those cities have documented health improvements, as noted on the following page.

There are two types of refuel systems: the fast-fill and the time-fill (an interesting euphemism for "slow"). Fast-fills are more expensive because they have an intermediate chamber to hold pressure between the compressor and the vehicle tank. Yet the time to fill is comparable to that of filling a gasoline tank. Time-fill systems take several hours, and the compressor is connected directly to the vehicle tank. This is the type used in home charging stations, which use domestically supplied natural gas as the feed. Honda partly owned a company named Phill that offered such a system. It recently went bankrupt, most likely because of slow uptake of the technology as a whole. It has resurfaced with different ownership. In the end, an overnight charge is not a prohibitive approach, particularly if commercial fast-fill stations are available for emergency situations.

Engine Type

The internal combustion engine for gasoline or diesel can accept CNG with very little modification. So, dual fuel supply is feasible and is in fact the case in most retrofits. But this means tolerating the combined weight and volume of a gasoline tank and a CNG tank. Also, conventional engines suffer about a 10 percent efficiency penalty using the gas. Dedicated engines can take advantage of the 125 octane rating of CNG. If the compression ratio is raised, more energy will be derived. The Honda Civic GX has a compression ratio of 12.5 to 1. This appears short of what is possible with the immense octane rating, but is likely a compromise on weight and cost. Also, some other combustion-related aspect might be limiting the compression ratio.

I have mentioned that fleet vehicles of all types are advantaged by CNG, as they would be with electric vehicles: charging/refueling infrastructure is simple. Public buses, delivery vans, postal vans, and so forth should find ready acceptance except for the retrofit time and cost. Fuel cost is low, and with shale gas the cost can be expected to be predictably low. Today reported figures are half the cost of gasoline per mile driven. This should be better with high-compression engines.

Long-haul transport cannot easily justify CNG because of its low energy density. Liquefied natural gas (LNG) is called for. This suffers a volumetric penalty of just 1.4, as opposed to 4.0 for CNG. However, refueling is much more problematic. In 2010, Volvo field-tested a truck using a 75/25 blend of LNG and diesel. Volvo claimed the range to be between 500 km and 1,000 km depending on conditions. In 2011, Volvo appears to have decided to go in the direction of DME. The company has announced a 13-liter truck engine running on DME alone for the 2015 model year, this applying also to Mack Trucks. For European long-haul, 1,000 km seems to be a comfortable distance to refuel. But again, LNG transport and dispensing is specialized. LNG for trains to replace diesel is also being piloted in India, the US, and elsewhere. Range is again an issue, and it will remain so until refueling infrastructure is more widespread. Recent advances in mini LNG production will be an important element. The key driver in India appears to be the cost of diesel rather than environmental gains.

Health Benefits of Substituting Diesel with CNG

The principal issue with diesel emissions is particulates. Engines and fuel have improved to the point that the old thick black exhaust is not really a factor any more. But the particulate loading is still high. An Italian study considered two scenarios in the period 2009 to 2011: a trend scenario of 3.3 percent replacement with CNG vehicles and an aggressive scenario with 10 percent replacement. In metropolitan areas they estimated a 1.3 percent reduction in deaths related to respiratory and heart diseases with the trend scenario and a 4 percent reduction in deaths under the more aggressive replacement scenario. Expected improvements included 4,197 and 13,115, respectively, fewer asthma attacks in adults. Similar figures were reported for lost work days and children's illnesses.

Such studies have also been conducted in New Delhi and Kuala Lumpur, in those cases with actual retrospective data. A New Delhi study (Akbar et al., 2005) claims 3,629 fewer deaths due to the switch to CNG by all public transport, which was ordered by the Supreme Court of India. This success has allowed expansion of the program to 22 major cities. Again, fleet operations are easier for implementation.

Europe is much more positively affected by the diesel switch to CNG than is North America, where diesel passenger vehicles are simply not in demand. That could change soon. Automobile manufacturers are realizing that the miles per gallon standard may most readily be achievable through partial fleet conversion to diesel with its 35 percent better mileage over gasoline.

Mandatory CNG for public vehicles (including taxis) and utility vehicles such as delivery vans in metropolitan areas, LNG for long-haul trucks, and possibly CNG for light-duty trucks: these are all worthy targets that could make a serious dent in our imported oil bill. Ultimately electric vehicles are the future, provided needed advances in battery cost and performance are made.

CHAPTER 18

Horses for Courses: Challenging the Orthodoxy in Fuels and Chemicals

"Did you ever have to make up your mind?"

—From "Did You Ever Have to Make Up Your Mind?" by The Lovin' Spoonful (written by John Sebastian)

Fuels and chemicals have been manufactured in the same way for a long time. The deeply ingrained beliefs centered on the economies of scale have caused plants to get larger over the years. Oil refineries used to be simple and as small as 10,000 bpd of output. The newest ones are topping a million barrels. Against this backdrop has arrived the concept of distributed production, with smaller refineries located closer to either the raw material source or the consumer. In this chapter I detail the advances and posit the notion that the methodologies will coexist, each suited to certain situations. Some horses run certain courses better. Kentucky Derby winners who have faltered at the slightly longer Belmont Stakes know the feeling well. In the fuels and chemicals area we are going to have options for the first time, but work remains to be done to realize this vision.

The concept of distributed production is not new. Electricity has embraced this to some degree when the technology was suited, such as with solar photovoltaics. The default option is still large generators exceeding 1 gigawatt, with transmission lines everywhere. While this has merit in countries such as the US, where transmission line losses average 7 percent, countries with losses up to 40 percent ought to consider alternatives. In these settings the vast majority of the losses are theft-related, but no matter the cause, it does not get there. This is a case where distributed production of electricity makes eminent sense. Fuels have analogs to this, as I discuss below; the thievery is more sophisticated and involves subversion of the subsidy system.

The early driver for innovating in the small footprint production space was stranded natural gas. Today roughly 5 trillion cubic feet of gas is flared (burned for no benefit) annually. This is one-fourth of the total annual consumption in the US and has a value of $20 billion, even at today's depressed prices. Most of the stranding is occasioned by the absence of pipelines to move the gas to market. In the case where the stranded gas is on an offshore platform, the space available for conversion to liquids is small. This waste was a major driver for the research.

More lately shale gas has provided a somewhat different justification. Unlike conventional wells, shale gas wells produce smaller quantities of gas per well. Shale gas wells average 1.5 MM standard cubic feet per day (scfd), compared with over 100 MM scfd for deepwater wells. The stranding in these cases could well be the absence of delivery pipelines. But if liquids could profitably be produced at the rig site or close to it, this would provide an alternative avenue for monetization over the construction of a pipeline. I refer to this as a virtual pipeline. Also, many wells are economically stranded, meaning the market price does not allow profitable production. In early 2014 this was the case for many wells in eastern Pennsylvania and the Haynesville play in Louisiana. Boom towns have become shadows of their former selves, if not outright ghost towns. The entire class of operations converting gas to liquids with some value I am dubbing "GTL Lite." In so doing I am defining "liquid" pretty broadly to include transport liquids as well as others, such as methanol.

The Case for Natural Gas as Raw Material

Natural gas is already the preferred feedstock for many chemicals, such as ammonia. But oil continues to be the dominant source for transport fuels and for certain chemicals such as ethylene and propylene. These last two are important precursors for much of what we wear or package. It has always been technically feasible to produce these from gas, but lately, with the spread between oil and gas prices, at least in North America, the economic viability has improved.

Figure 14 shows the selling price per MM Btu for several important products in a snapshot taken in April 2013. Market conditions can cause variations, but by and large the numbers for 2014 are not far off the ones shown. In early 2014 the exception was methanol; its price had a large increase, but it is likely temporary. The interesting point of note is that when compared on this uniform basis of energy content, the prices of the fuels are in parity with those of chemicals. These do not share the same markets, so why would this be so? My explanation is that they are all sourced from oil, which has output strictly controlled by the OPEC cartel. When some of these are sourced from natural gas, some separation could start occurring.

Figure 14. Value of different fluids produced from natural gas

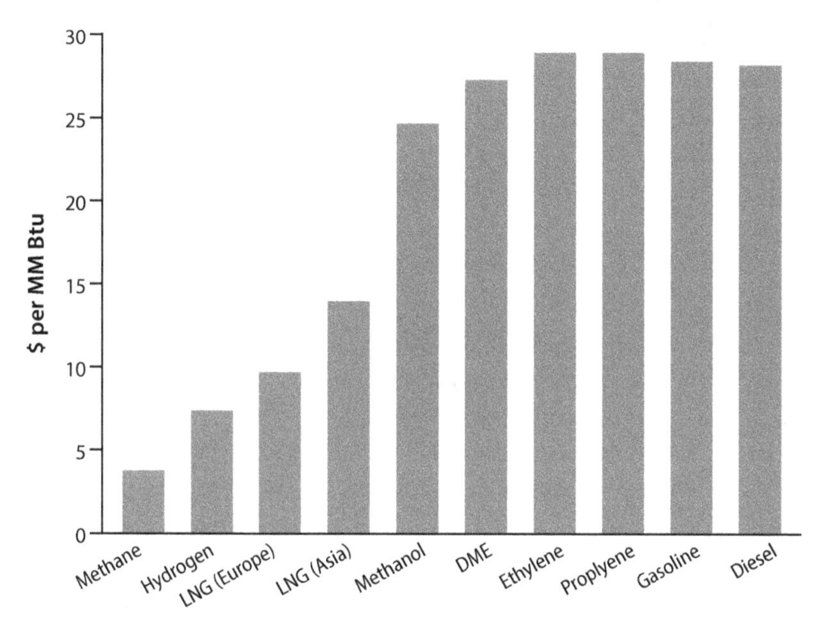

Source: Figure courtesy of RTI International.

If oil were to be priced at $100 per barrel, it would show up at $18/MM Btu on the figure. With some processing cost, it could be refined to one of the fuels and chemicals shown at $27/MM Btu or so. But the same chemicals could be synthesized from natural gas starting at $4/MM Btu. Sure, synthesizing the chemicals from natural gas would incur higher processing costs, but the margins would still likely be much better. In the following chapter, I go into more detail on methanol and point out that it could be made for around $12/MM Btu. That leaves a lot of headroom to the selling price.

Economies of Mass Production

A school of thought is developing that economies of mass production could displace economies of scale in many instances. All large-scale chemical plants, and in this I am lumping fuel production plants such as refineries, have in common the method of assembly and construction. Each is custom-designed even though many elements may be commonly used elsewhere. Then it is assembled and tested on location by skilled personnel. This inevitably constrains the locations to those in which skilled labor is available. Certain areas such as the Houston Ship Channel and the New Jersey corridor are advantaged and form areas of concentration.

The new concept would entail building small reactors that are economical in their own right. Part of the objective would be to design them to be manufactured in volume and enjoy the economies of mass production. These could be deployed singly in distributed fashion. They could also be collected in one location and hooked up together. This process would be much simpler because it involves a much lower level of complexity than a custom plant does. Another important consideration is that they be low cost and not necessarily long-lived. One would sacrifice long life for low cost and low maintenance. An analog is the internal combustion engine: relatively low cost for the horsepower and economical for the intended lifetime. By contrast, large chemical plants are designed to last decades. Aside from the high cost to finance, the plan has to ensure raw material supply for long periods. This uncertainty makes financing large plants more expensive through a higher discount rate.

A large plant composed of small modules hooked together would also rely on another feature. It would have to be highly automated, with sensors to detect the performance of each module. Since these modules would be designed to be replaced as frequently as every five years, monitoring would be a key to success. Additional economies could be derived from remote control of these modules, thus minimizing the need for skilled operators at the location. This technique has been used to advantage in the oil and gas business in exploiting hydrocarbons in remote and otherwise challenging areas such as ultra-deep water. Experts can monitor and control multiple locations, reducing cost and providing job satisfaction through not having to spend time in the field.

Making Choices

Until the vision described above is realized, there really is no choice to be made. Small distributed reactors not yet taking advantage of the economies of mass production exist now. This is either already here or right around the corner, depending upon whom you believe. At least six companies have natural gas to diesel, methanol, or DME processes in late stages of commercialization. Each is about 100 times smaller than conventional plants.

Figure 15 illustrates the logic companies might follow in choosing which type of process or plant to build in a particular situation.

For large sources of high-quality gas, be it methane or ethane, it makes sense to have a conventional-size conversion plant located in the vicinity. At the other end of the spectrum, is the situation where the field is small and the gas of poor quality. Dry gas in smaller accumulations would fit in that

Figure 15. Logic for choosing a monetization solution

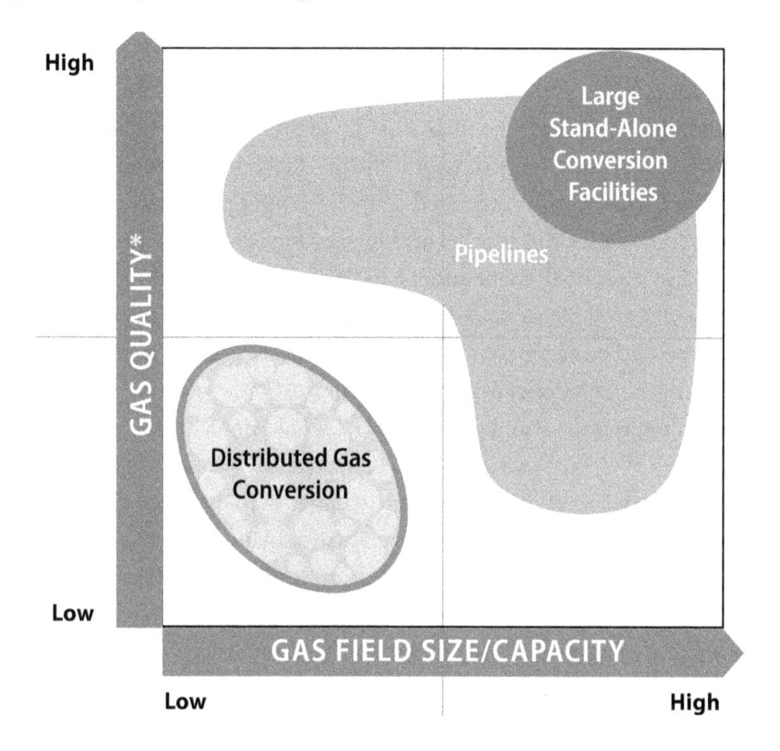

* Gas-field location, gas wetness, impurities, etc.

category. They would be well-suited to small-footprint plants distributed to be close to the source. Intermediate locations on the chart could be served by combinations of modules or by pipelines transporting to existing conventional facilities. Such logic will drive the best mix of conventional and distributed production. When systems taking advantage of economies of mass production come on stream, they will compete with conventional technology for new build capacity.

The natural gas price disparity between the US and Canada and the rest of the world is driving a manufacturing renaissance in the US. This added capacity, often with foreign investment, will force investors to choose between technologies pretty soon. Assuming the new approach is viable, the country would benefit if distributed production became available before too much new capacity comes on stream.

The Need for New Business Models

GTL Lite technology appears to be well on its way to becoming available. Especially in the case of monetization of stranded gas, however, the hurdles go beyond those of technology incorporation. Security of supply is much simpler with small units, particularly because a lot of the producers are themselves small. Further, it is immensely simpler if the producer and conversion process owner are vertically integrated. Even long-term pricing would not be a contractual hurdle. But true vertical integration is hampered by the fact the gas producer has no domain understanding of the other area, which is essentially a chemical plant, albeit a small one. A construct which could work is a partnership for the sole purposes of that one venture. But the lack of domain understanding could hamper even partnerships.

In theory, integrated oil and gas companies would not face this hurdle; in practice, however, the separation of management of the down and up streams could be a complication. Also, the competitive advantage of the large, integrated companies rests in part in the ability to finance and manage giant projects. In GTL Lite projects, they would be on similar competitive footing with the small guys. In fact all the players in the GTL Lite space right now are small outfits. This brings to mind an important distinction between the regular and Lite versions. Large-scale facilities can be designed and built by only a very small handful of companies; fewer than five would be entrusted with a world-scale GTL plant. Lite plants would not only open the field to smaller oil and gas players, but a large number of companies, possibly a hundred, could perform the engineering and build. This is good for the industry as a whole.

Distributed Production in Remote Villages

Over half the population of India has access to little to no electricity. What meager power they get is from diesel generators, sometimes kerosene. The latter is also used for lighting, as is paraffin. Cooking is done with biomass in open stoves or sometimes with kerosene. All but the biomass is derived from oil, the majority of which is imported at world market prices. Diesel and kerosene are heavily subsidized. Kerosene in particular has the subsidy applied early in the distribution chain and not at the consumer. Such subsidies are subject to a lot of corruption. In any case, government-subsidized diesel and kerosene constitute a burden on the taxpayer, and economical substitutes would make a difference by providing cost-effective energy without the need for heavy subsidies.

Biogas produced from organic wastes could be converted to a liquid fuel using methods discussed earlier in this chapter, with the small-scale methods likely being the most viable. A technology in late-stage development (Dayton, 2013) processes woody biomass into a synthetic oil. This oil has been successfully used as a refinery feed. In this instance, though, it could be used directly for electric power generation in a Stirling engine, which is an external combustion engine suited to taking fuel other than the conventional diesel. Another use for this oil could be in cook stoves. Direct use may be possible, but more likely it would need to be converted to a kerosene-like fuel. This would require a relatively simple step known as hydrotreatment.

Replacement of both diesel and kerosene will have positive impacts upon respiratory and cardiac health. Diesel generators emit particulate matter, the most noxious being $PM_{2.5}$, which are particles smaller than 2.5 microns (micro meters) in size. The issues surrounding this are described in the box.

Particulates Matter

Particulate matter emitted from combustion systems constitutes a significant health hazard. Particles smaller than 10 microns are known as PM_{10} and are associated with some respiratory illnesses. $PM_{2.5}$ is much more hazardous because it penetrates tissue. It is implicated in cardiac and respiratory illnesses. The principal sources of exposure for the general public are coal combustion and diesel engine exhaust. Advances in filters continue to improve the situation with respect to diesel engines but particulate matter is still a hazard. Some estimates (Muller & Muller, 2013) peg the annual unintended US mortality at 70,000; morbidity numbers are more difficult to estimate. The mortality number for the world is close to 1.5 million.

In the US, progressive replacement of coal by cheap natural gas will do its part in attacking the problem. This is likely to be especially effective because the plants being replaced are the oldest ones and therefore the worst actors in regard to $PM_{2.5}$. Diesel displacement is also likely to occur, as discussed in chapter 26.

But in growing India and China, coal will continue to be the major electricity source. In India and parts of Africa such as Uganda, the problem ought to be attacked at the village level. Here the electricity generated is from diesel, and cooking is done using kerosene or woody biomass. All of these generate copious $PM_{2.5}$. In the case of cooking, the portion of the population most at risk are the women and the small children who by custom are in close proximity. Substitutes for these fuels—solar electricity and innovations such as smokeless cook stoves—would do much to improve the lot of these unfortunate billion people.

When introducing innovation in village settings, technology is necessary but not sufficient. Sufficiency requires a depth of understanding of the societal context. If you build it they may not necessarily come. For example, if a kerosene substitute were to be found, the folks in the kerosene supply chain are likely to place roadblocks. One smokeless cook stove introduction was hampered by the villagers' simply not wanting to use them because the appearance was not acceptable. The poor, too, have aesthetic sensibilities. These hurdles can be traversed but need to be addressed right up front using local knowledge.

Small-footprint chemical conversion processes present a challenge to the orthodoxy in the fuels and chemical industries. Certainly they allow for distributed production of fuel. They even offer the promise of substitution of economies of scale with economies of mass production. In the end the method most fit for the purpose will be adopted and the new and the old will coexist.

CHAPTER 19

Advantage Methanol

"Mere alcohol doesn't thrill me at all. /
So tell me why should it be true"

—From "I Get a Kick Out of You" by Ethel Merman (written by Cole Porter)

As I have discussed in previous chapters, a cheap natural gas future is conducive to the production of liquids for internal combustion engines. Fischer-Tropsch synthesis gives us drop-in replacement for oil-derived gasoline or diesel. The methanol-to-gasoline (MTG) process likewise produces moderate octane gasoline, with methanol and dimethyl ether (DME) as intermediate products. Methanol can be processed to DME, which can be a cleaner-burning substitute for diesel. But what of methanol itself? Could the supporting actor become the marquee attraction?

A careful examination shows that if certain measures are taken, methanol could be a viable endpoint. In fact, of all the gasoline substitutes derived from natural gas, methanol may well have the inside track based solely on economics. In tennis terminology, methanol has moved from the deuce to the advantage court. That is one step from winning the game. Harnessing the high octane rating would give it set and match against the perennial champion gasoline.

Widespread acceptance of any gasoline or diesel substitute requires three conditions to be met. The first is low cost per mile driven relative to the incumbent fuel, preferably without the assistance of subsidies. In the event of subsidies, the consumer would need some assurance that the post-subsidy price would still be attractive. In fact this assurance is likely more needed by the retailer who would need the security of long-term demand to incur the expense of a dispensing system. This brings us to the second condition, the need for a low-cost distribution infrastructure, including pumps at retailing stations within reasonable driving distances for most consumers. This last was one factor in the unpopularity of E85 (Kindy & Keating, 2008). The final condition required is that of easy availability of vehicles that are able to operate with the alternative fuel. In the limit, all new vehicles ought to be able to run on any mix of gasoline and alcohol alternatives. Bills in both chambers of Congress await passage that would ensure that most new vehicles after a period of time would be flex-fuel vehicles (FFVs).

Price of Methanol Compared to Gasoline

In early 2012, the methanol wholesale price was $1.13 per gallon. I will use this figure rather than the unusual $1.50 prevalent in early 2014, because the lower figure is more representative of a steady state case. Distribution and markup will add 10 cents, and another 20 cents for federal and state taxes brings the total to $1.43. To compare with gasoline, this needs to be doubled because methanol has about half the calories of gasoline. Later I will discuss how this calorific disadvantage can be ameliorated or eliminated. But in the normal situation for use in an FFV, the effective cost of methanol is $2.86. On the day of this estimate, regular gasoline was selling for $3.43 as a national average. So, clearly methanol enjoys a cost advantage. Add to that another fact: methanol has an octane rating of 117. So the comparison ought to be with high-octane, or super, gasoline, not regular. In most states one would need to add another 25 cents, bringing the gasoline total to about $4.04. As a practical matter the most likely fuel used would be M85, comprising 85 percent methanol. But the substitution comparison based on the neat liquid is still valid.

All this raises the question: will methanol prices in the future enjoy this advantage? Since the bulk of methanol today is produced from natural gas, the question shifts to the forecast in natural gas prices. In fact, at a methanol policy forum in 2011, David Sandalow, Assistant Secretary of Energy for Policy and International Affairs, in touting the value of methanol, cautioned, "If natural gas prices increase substantially and price differentials between methanol and gasoline revert back to historical norms, the economics could be difficult." This is precisely the concern addressed in the analysis below. I conclude that methanol will be cheaper than gasoline on a per-mile basis for decades. The only wild card is extraordinary demand for natural gas outstripping the production capability

The cost to produce methanol is plotted in Figure 16 as a function of natural gas price. In March 2012 natural gas prices were essentially at decadal lows, driven by an abnormally warm winter. These prices are not normal; the October 2010 figure shown is more what one could have expected. That figure, incidentally, was roughly the average for 2010 as well. One readily observes that at those values, methanol can be produced for about 50 cents per US gallon.

Figure 16. Methanol cost as a function of natural gas price

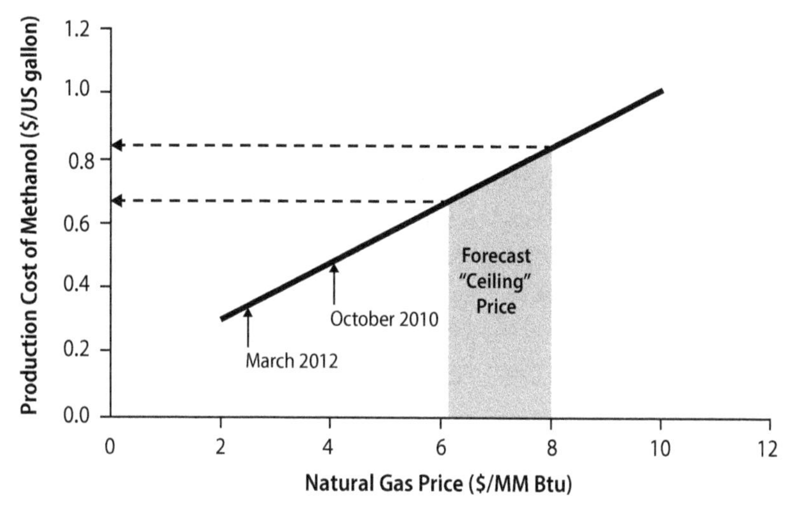

Source: Figure courtesy RTI International

Keeping in mind this is cost to produce and adding a profit of 15 percent gives about 58 cents. Add to that the aforementioned retailing and tax components and you get about 88 cents per gallon.

In chapter 3, I presented a model indicating a ceiling for gas price between $6.50 and $8.00 per MM Btu. Support for this is available from the work of Amy Jaffe and colleagues (Medlock et al., 2011), who used different scenarios of shale gas development. In none of their scenarios do they exceed decadal averages of $6.50 for the next three decades. More recently, the EIA forecast calls for the price to stay below $5.50 out to 2025.

My forecasted ceiling is plotted in Figure 16 and represents the upper end of what one could expect for the methanol cost of production. At the highest end, methanol rings up at a pump cost of about $1.28 per gallon (85 cents to produce plus 15 percent profit plus 30 cents in distribution and taxes). An important point of note is that these calculations notwithstanding, the price of methanol will be driven by market factors. However, as one can see there is a fair bit of headroom between gasoline and methanol, and even accounting for variability in both, methanol is still a viable choice. Gasoline price will always be driven by world events, whereas methanol will be largely regional. In that sense one could expect more stability. Natural gas represents about 75 percent of the cost of methanol production, so stability in that commodity will at least keep cost under control.

Another factor is that methanol can be produced from coal and biomass, so alternative feedstocks are a moderating element. In particular, low-grade coal such as lignite is a useful feedstock. Lignite, available at the mine mouth for $25/tonne (recent figure from the EIA) will yield methanol at a cost of around 70 cents per gallon. So, if my forecast is vitiated by unusual demand for natural gas, plentiful low-grade coal can kick in. This coal is not particularly useful for electricity production because of its high moisture and ash. In fact, the high proportion of this useless component adding to freight cost is the reason I advocate mine mouth processing. With vast deposits in Texas and states northeast from there, the mine mouth is not far from infrastructure. Of note regarding the switch to another raw material is that the process after the formation of syngas is the same regardless of the feed character.

Methanol Production Cost

For this computation I used a standard plant producing 5,000 tonnes per day of methanol. This translates to about 40,000 barrels per day. As a frame of reference, the announced Sasol GTL plant for Louisiana is rated at a bit over double that figure. In general methanol plants require less capital than F-T process plants. The 5,000-tonne-per-day plants could be expected to cost about $800 million, whereas an F-T plant of the same capacity would be about two and a half times that. The time to construct would be similar in ratio.

This plant would consume 150 MM cfd per day of natural gas. This is the output from about 100 Marcellus wells, or about 4 to 7 pads. Such plants could be distributed in the producing areas, keeping the jobs local. In fact, the simplicity of the process could well allow even smaller plants without much loss in economies of scale.

Viability of Flexible-Fuel Vehicles

The intent of an FFV would be to use any mix of gasoline, ethanol, or methanol. While there is some difference of opinion, most believe that the current FFVs could use methanol. Possibly a somewhat more robust fluorinated elastomer would be needed. Certainly the software or firmware would have to be modified to allow for a methanol mix. With an M85-filled tank, the range loss would be about 42 percent. In other words, a car with a range of 350 miles would now go about 200 miles with a full tank of fuel. The consumer would have to decide whether a lower price and lower emissions are worth filling up more frequently.

Today that lower price would be calculated as follows: 15 percent at an average gasoline price as noted above of $3.43 plus 85 percent at $1.76 (natural gas at $4 delivers methanol at this price after the doubling to take into account the fact gasoline has twice the energy content) equals $2.06 for a gallon of M85. I used the price of regular gasoline in the calculation. But M85 performance would need to be compared to high-octane super, priced at $4. Still, the consumer would pay half for the fuel in exchange for filling up almost twice as often. Eventually, if auto makers make larger tanks to accommodate this, then the fill up will be at a normal frequency. Most of the public would take that tradeoff; these are folks who are inclined to drive a mile or more for 10-cents-cheaper fuel. The comparison to the price of super gasoline is not completely fair because regular-compression engines get no benefit from the higher octane rating. The clever nomenclature "super" unintentionally causes some of the public to erroneously believe it is better.

The distribution argument goes as follows. If a large proportion of cars were to be FFVs, and if methanol had enough allure to get some fraction of these to owners to use it, a dedicated M85 pump would pay for itself. Furthermore, all the other cars on the road have either regular compression, needing 87 octane regular gasoline, or sporty cars with higher compression. These latter engines need 91 octane fuels. Cars currently in service do not need a third grade of gasoline. In fact the practice today of having octane ratings of 87, 89, and 93 makes no sense. Future pumps would be 87 and 91 octane gasoline and M85.

High-Compression FFVs

This is the future: smaller, more powerful vehicles with a longer range and, in the case of methanol, nearly half the carbon emissions for the same miles driven as with gasoline. The "Flex-Fuel Fairy Tale" (chapter 25) lightheartedly alludes to such things. But fantastic sounding though it may be, there is firm scientific basis for these assertions. All of it relies primarily on one feature: all three gasoline alternatives—methane, methanol, and ethanol—have in common the feature of extremely high octane ratings.

Our goal: an FFV that accepts any alcohol combination with gasoline, and also methane if practical. Of course the gasoline portion would need to be small or zero for the octane number to be high enough. Certainly E85, M85, and CNG would function in such an engine. But until there is a breakthrough in ethanol production cost and a similar advance in storing CNG in a smaller volume than currently, methanol will be the game. So, any further discussion will be for this fuel. But an FFV accepting both alcohols, and possibly methane, has the virtue of providing consumer choice.

How Gasoline Is Affected

Gasoline appears to be a loser on this, and some may argue that it would be premature to force a wholesale switch. As I discuss below, there is a variant that allows a mixture with a majority of gasoline in an effectively high-compression engine, known as the direct injection engine. Also, existing vehicles will be on the road for a long time. But if gasoline is forced into the position of being just another fuel, the future sought by Gal Luft and Anne Korin in *Turning Oil Into Salt* (Luft & Korin, 2009) will be realized.

Their thesis is that salt used to be a strategic commodity because of its critical function in preserving food. Wars were fought over it, and people were paid in it. The word *salary* comes from salt. Then refrigeration changed all of that. Salt became a useful, even essential, but not strategic commodity. Luft and Korin suggest that cars allowing fuel choice will render gasoline, and by extension oil, a nonstrategic commodity.

First let's discuss the direct-injected, alcohol-boosted engine. This is a relatively high-compression engine primarily powered by gasoline. When the likelihood of premature ignition is detected, methanol is injected using a special line and port. It has a high latent heat of evaporation, so when it is injected, it cools the chamber. This suppresses premature ignition, eliminating the knocking and realizing the high efficiency of the high compression with very little total methanol. Such a gasoline engine can give efficiencies equal to or greater than a turbocharged diesel, which is more expensive.

The greater promise from the standpoint of displacing gasoline is an engine with a much higher compression ratio, closer to 17. Any alcohol blend with low gasoline content would operate in such an engine. Work at the EPA laboratories in Detroit (Brusstar et al., 2002) and at MIT (Bromberg & Cohn, 2008) has shown that both M85 and E85 in a low-displacement, spark-ignited, high-compression gasoline engine can obtain efficiencies exceeding that of diesel engines. So, in effect not only does this eliminate the near-factor-of-2 calorific (mileage) penalty of methanol, but you end up with a smaller engine that gets better mileage than a gasoline equivalent. Also a higher-torque, sportier car to boot!

Heavy-Duty Vehicles

This class of vehicle uses of 2.5 MM bpd of oil, out of a total 8.1 million barrels imported each day. Consequently it is a logical target. FedEx recently announced an intention to switch its trucks to LNG. MIT research has shown (Cohn, 2012) that a direct-injected spark-ignition engine can be made smaller and more efficient. A typical 15-liter-displacement engine can be replaced with a 9-liter engine. This utilizes the high octane rating of methanol as well as the evaporative cooling mentioned earlier. The engine weight can be reduced from 3,400 pounds to 1,800 pounds, and the exhaust control can be simplified. However, the fuel tank needs to be 300 gallons instead of 200 gallons. But the added weight of this feature is completely offset by the lighter engine.

The High-Efficiency FFV: How Do We Get There?

Any large industry such as the automotive industry is slow to change, and with good reason. The original ethanol-based FFV was assisted with a federal subsidy to cover the engine modifications. If the broad mandate for FFV capability on most new cars goes through and if methanol production gears up, the compelling economics will cause consumer demand. This will create fueling infrastructure, perhaps assisted by one-time subsidies. At this point the commercial risk of introduction of a high-efficiency FFV is minimal. By all accounts the automobile cost will be lower than for the gasoline counterpart of equivalent performance. This is in part due to the smaller engine and the more inexpensive emissions control equipment.

The automobiles will appeal to the buying public for their defining characteristics of a small, high-performance engine with long range fueled by a cheaper alternative to gasoline. The feature of nearly half the carbon dioxide emissions per mile as compared to an equivalent gasoline engine will also appeal to some. This last comes about from the fact that methanol has about half the carbon of gasoline, which causes the mileage penalty in an ordinary engine. But the high compression feature simply makes the engine more efficient, in effect negating the low carbon penalty.

An interesting entrée could well be through the military. On the one hand, they should have an interest in more efficient fuel utilization to minimize fuel transported to the front lines. On the other, methanol production at major bases using the feedstock of convenience would not be out of the question. A civilian version of a military vehicle could then follow, much as happened with the Humvee (originally the High Mobility Multipurpose Wheeled Vehicle, or

HMMWV). Ironically, the high-efficiency FFV (Heffvee) would be almost the opposite of the gas-guzzling Humvee.

When Nobel Laureate George Olah and colleagues proposed the Methanol Economy (Olah et al., 2009) first in 2006, they were a bit ahead of their time. Shale gas had not made its presence felt to assure a long-term future of moderately priced gas. Now the promise can be realized. While methanol will be the main driver in gasoline replacement, uniform acceptance of first FFVs and then Heffvees will allow for ethanol to be used as well. Gasoline will take its place as just another fuel, not the dominant one.

The Road to Energy Independence

Responsible production of shale gas will essentially eliminate import of natural gas. That leaves the big ticket item—oil. Here, too, the notion of independence can usefully be defined as independence from distant and unreliable sources. The first step could be to target the oil passing through the Strait of Hormuz.

In 2012, the EIA forecast that in 2022 we would import 7 MM bpd, down from the 8.1 MM bpd in 2011 (EIA, January 23, 2012). I think that if pipelines are built from North Dakota, oil from the Bakken formation will eat into this number more than already forecast. In fact, the 2013 forecast called for import reduction to 6 MM bpd by 2018. Going with the newer figure, first subtract imports from Canada and Mexico. Canada can be expected to ramp up its current flow of 2.2 MM bpd to at least 3.0 MM bpd. We have a special relationship with the Canadians: the bulk of their oil can only be refined in the US. Aside from the high carbon footprint of this oil, this is a desirable and secure relationship. Mexico currently supplies 0.8 MM bpd. This is at considerable risk of decrease but we will leave it at that figure for 2022. This, too, is heavy oil suited to our refineries.

Of the 2.2 MM bpd balance, I estimate about 1.7 MM bpd passing through the Strait of Hormuz. Therefore, one strategy would be to target oil alternatives to that level. Ignoring for the present the fact that a barrel of oil does not generate a full barrel of transport fuel, we can target 1.7 MM bpd of oil replacement. A rough calculation of all sources indicates this is viable, as enumerated below:

- Sasol has already announced construction of a GTL plant in Louisiana reportedly rated at 96,000 bpd of fuel. Assume at least one other such, bringing the total close to 200,000 bpd from GTL emboldened by low gas prices. (Shell's cancelling its Louisiana GTL plant in early 2014 is seen as unique to their company, not necessarily a trend.)

- In my chapter on Alaska (chapter 15), I suggested means by which at least 200,000 bpd capacity could be added to the pipeline. Absent some such action, the 500,000 bpd currently sent down from Alaska is at risk due to pipeline economics.
- Long-haul trucks switching to LNG or methanol could reasonably target 20 percent of current fuel usage, which accounts for 0.5 MM bpd of oil.
- Methanol, ethanol, CNG, biofuel, and electric cars could target 1.0 MM bpd. A significant part of this, and relatively straightforward, would be CNG displacement of diesel in metropolitan public and commercial transport.

An angle other than a Strait strategy is a study of the marginal domestic barrel and what it replaces. New domestic oil production is all light sweet oil. This is most like the oil from Saudi Arabia and Nigeria. So that may make sense as the first to be displaced, and in fact in 2014 Nigerian imports are down to a dribble. The Saudi portion is, of course, Strait-related and currently stands at about 1 MM bpd. Similarly, the uptick in Canadian oil that I predict will displace heavy crude such as that from Venezuela is currently about 0.65 MM bpd. The main point is that crude quality is variable and refineries are choosy, so country strategies have to recognize this.

Shale gas produced responsibly will be a key enabler for methanol to be produced at prices attractive with respect to gasoline. Broad availability of FFVs and associated fueling infrastructure will give the public choice; more on that in a later chapter. Tomorrow that choice could include other alcohols or methane, and a suggested high-performance FFV will enable that. Today methanol appears to be particularly advantaged. Ultimately, gasoline (and diesel) can be rendered just another player, not a champion. Game, set, and match.

PART IV

Informing on Policy

IV. Informing on Policy

The societal impact of shale oil and gas production has been profound. It has informed local, state, and national policies.

Shale gas replacing coal for electricity generation is credited with the US reverting to 1992 levels for carbon dioxide in the atmosphere, the best performance of any country. Yet without doubt, the low price of gas will have a chilling effect on renewable energy on pure economic grounds. The shale oil–mediated plummet in oil price will also affect transport fuels. Oil-based transportation fuels will be displaced by substitutes such as natural gas, woody biomass, and oil derived from seeds. Policy measures will be necessary to level those particular playing fields.

In this section I explain energy level and octane of gasoline alternatives such as ethanol, methanol, and natural gas. A light-hearted treatment with sound technical underpinnings is in the Flex-Fuel Fairy Tale chapter (chapter 25).

The drop in oil and gas prices worldwide has diminished the treasuries of the producers, reducing their ability to impose political will using oil or gas supply shutoff as a weapon. This has national security implications, not merely energy security ones.

CHAPTER 20

Turning the Pennsylvania Two-(Mis)Step Into a Waltz

"There should be sunshine after rain"

—From "Why Worry" by Dire Straits (Written by Mark Knopfler)

There is an old adage: the people in the front get shot. Pennsylvania took some bold steps in the development of shale gas. They allowed shale gas production and the associated fracturing process. New York put a moratorium on fracturing and decided to study the matter. France and Switzerland did much the same. An industry-sponsored study estimated that in 2012 Pennsylvania added $14.0 billion in economic activity, generated $3.0 billion in additional state and local taxes, and supported nearly 100,000 jobs (Considine et al., 2011). Industry opponents argue that groundwater was polluted. While each of these figures and assertions can be contested, and they are, they have underlying truths.

Fracturing operations have been routine for 60 years. Fracturing-enabled production of gas, and to a lesser extent oil, has been commonplace in Texas, Louisiana, Arkansas, Oklahoma, Colorado, Wyoming, Montana, and the Dakotas. So why the fuss about Pennsylvania, the birthplace of oil in this country?

The origins of oil in the US notwithstanding, Pennsylvania is largely unused to this sort of activity. The prospects lying in populated areas were also a factor. But the biggest impact likely came from the fact that the safe and cheap disposal of fracturing water was not practical here. In those other states, EPA-certified UIC Class II deep disposal wells were commonly available. The geology of Pennsylvania militated against this. In some cases the measures taken were woefully inadequate, such as sending the contaminated water to municipal disposal sites, which were ill-equipped to handle this waste. Others allegedly dumped the water. Yet others paid a lot to haul the water to deep disposal sites in Ohio.

Making a Virtue of Being Late

This statement has the makings of an oxymoron. In many settings it certainly is. For example there can be no discernable virtue in being late for your own nuptials. Being late for one's own funeral, if that could be pulled off, has decided good points.

Being late is not the same as coming in second. Nobody knows that Tom Bourdillon and Charles Evans were within 300 feet of the summit of Mount Everest three days before the second team of Edmund Hillary and Tenzing Norgay got to the top. Bourdillon and Evans likely did not even make it into Trivial Pursuit.

In the business of innovation there is a body of literature on the value of being first. "First mover advantage" is firmly in the business lexicon. But so is the "fast follower" principle. Indubitably, fast followers could be faced with patents preventing their success. Intel went out in front early and never was materially threatened by Advanced Micro Devices (AMD). But many businesses have been built on the premise of letting somebody else develop the market and make the mistakes.

What does all of this have to do with energy? The history of development of shale gas is instructive. After the realization that horizontal wells and fracturing enabled gas production from these tight rocks, the early attempts employed methods previously used. In particular these involved using sugars as thickening agents to more easily fracture the rock. The sugar residue impaired production. Newer techniques, in areas such as the Marcellus, use "slickwater." The results have been dramatic, albeit at the expense of higher volumes of fresh water.

All of the foregoing is just plain building on the experience of the past. This essay on the virtue of being late keys on the point that if fate has dealt you a hand that causes you to be late to the party, find ways to make that a positive. This is the opportunity presented to the areas of the East Coast that have not yet materially been swept up by the shale gale. These include Ohio, West Virginia, Maryland, and North Carolina. These states must institute measures whereby the exploitation of the resource is done in an environmentally sound fashion while still maximizing the realization of economic value for the communities affected. The Grand Experiment in Pennsylvania will be highly instructive.

The state is now taking stock of the substantial pluses and minuses. The actions will be watched closely by late-to-enter states such as West Virginia and Ohio. The ones on the sidelines (such as New York, Maryland, and North Carolina) will also watch with interest. Sometimes it pays to be a fast follower.

Taxing Natural Gas Production

We all recollect the now-famous phrase spoken by then-presidential candidate George H. W. Bush: "Read my lips: no new taxes." The public simply abhors new taxes and the politicians who propose them. House Bill 1950, the Marcellus Shale legislation, was passed by the Pennsylvania General Assembly. The major provision was an "impact fee" to be levied by each county. It would be proportional to the market price of the gas and have a time limit of 15 years. Understandably, the governor is loath to call it a tax.

While this appears to smack of politics-laden semantics, the suggested construct makes impact fee the accurate term. The levy is made by the counties and the proceeds are shared by the local municipality, the county, and the state. This allows for the money to be used explicitly to offset negative impacts of the industry, such as road damage. If targeted exclusively to this, not only would the term be completely accurate, but also the levy fair with respect to the industry. The state share may well be used in part to further job-related programs.

The public is more likely to accept gas production if it realizes that some of the proceeds are being used to benefit the community. This is especially the case for the majority who do not profit from lease royalties and yet suffer the traffic and other production-related inconvenience. In some ways this impact fee would stand in stark contrast to how the externalities with respect to coal and imported oil are largely given short shrift. In North Carolina, the Mining and Energy Commission, which I chair, has no direct jurisdiction on the taxation point but has been asked to study the matter. The Commission recommended to the state legislature that an impact fee be assessed, distinct from a severance tax, with the impact fee proportional to the number of fracturing stages. This metric more precisely relates to road and bridge damage because during those stages there are bursts of truck activity for the transport of water and sand in the main. The Commission urged that the bulk of the impact fee go to local municipalities.

Pennsylvania is the only major gas-producing state without a severance tax on hydrocarbons. This is basically a tax on either the volume of gas or its value and often has limited or total forgiveness for certain situations. As in the case of Pennsylvania's suggested impact fee, these tend to have sliding scales based on gas value.

Environmental Protection Measures

The report of the Pennsylvania governor's Marcellus Shale Advisory Commission is also instructive to any other states wishing to learn from the Pennsylvania experience. It covers a lot of ground, some of which is reflected in the discussion in chapter 30, "Policy Directions." The resulting bills passed in Pennsylvania fairly faithfully follow the recommendations of the Commission.

An important provision is the requirement to have baseline testing of proximal water wells prior to drilling activity. In addition to requiring this, the report stipulates that the operating company perform the task. They are also required to use an independent testing laboratory that is pre-certified by the state. The Pennsylvania Department of Environmental Protection (DEP) posted a list of state-certified testing labs to facilitate the process. A variant of this process could be a requirement for notification of proximal homeowners when a drilling permit is granted (this is in the DEP's recommendation now) and county assistance for proximal homeowners to get the baseline well testing done at the expense of the operating company. Among the advantages of this method would be the direct involvement of the homeowner in both selecting the testing entity and keeping track of the results. In North Carolina we have drawn up rules pretty much along the same lines. We have added the feature of periodic testing following the drilling. Other states, and eventually countries, would be advised to do the same.

An interesting recommendation is to double the civil monetary penalties for violations of Pennsylvania's Oil and Gas Act. The DEP is to be given authority to act in this matter and will generally be given a bigger stick for noncompliance, including suspension or revocation of permits. In general, so many of the needed regulations on production-related care and diligence are just good business for the operator. Punitive action ought to rarely come into play. This should reduce the cost of enforcement. Also reducing the cost would be the effective use of the latest technology in monitoring and transmission of the data to central locations at the oil and gas company. Remote monitoring and control of well operations are coming into vogue and would facilitate this regulatory purpose. This would largely negate the need for DEP physical presence on the rigs, avoiding a good deal of cost in the process.

Workforce Development

The state of Pennsylvania recognizes the need for thousands of trained workers in coming years. This is proving to be something of a challenge in some communities. The companies providing drilling and completion services provide their own specialized training. But base levels of qualifications are required depending upon the job. For many jobs, the required training can be as little as an associate's degree. Others require a bachelor's degree, preferably in technical disciplines. But almost all jobs face some common features: the hours are long (often unpredictable) and the work is hard. Furthermore, workers trained by these companies may be required to operate in other states and sometimes other countries. While the employers will be motivated to keep the personnel close to their original homes, the nature of the oil and gas business is one of a commuting life in many instances.

The Cranberry Effect

Cranberry Township in Pennsylvania has seen immense revenue growth due to the shale gas boom. No shale gas wells have been permitted or drilled in the entire township. What they did was create an atmosphere that attracted shale gas players, which located regional offices there. Support personnel such as accountants, lawyers, repair and maintenance outfits, and other professionals followed. It was simply a great place to live, work, and be entertained.

In some ways earlier history with another industry gave Cranberry an advantage. The town had succeeded in attracting a major research and development complex of Westinghouse, with nearly 5,000 employees. To do this, they developed the type of support services attractive to moderate- to high-income families. When shale gas drilling commenced all around them, they successfully targeted the operating entities. Today, the regional branches of several of the players in the Marcellus are housed in Cranberry.

This ability to get the gain without the pain I dub the "Cranberry Effect." They could in fact take the Effect to the next level by encouraging the downstream processing to be done in the township. One target could be the synthesis of anhydrous ammonia from methane. As noted in chapter 13, when methane is cheap this is a high-margin business. The other would be relatively small-footprint ethylene crackers along the lines suggested in chapter 14, "The Ethane Dilemma."

The town's close proximity to the ethane from wet gas would make the location particularly attractive. These jobs are high paying and tend to have well-scheduled shifts, unlike the drilling business. The work would also not be considered hard in comparison with other industrial jobs (some of the reluctant drilling job applicants had given this as a reason for recalcitrance). Most importantly, workers would be hired for that location only and not be required to spend time in other states. In other words, none of the objections to the drilling jobs would be applicable here.

The unusual time patterns can be a severe handicap to family life. It is less so for young workers without families, and seniority usually brings with it more regular hours. Studies among uranium miners in West Virginia have identified behavioral problems caused by the nonstandard working hours. In some cases the incidence of alcoholism went up.

In Butler County, unemployment numbers in 2010 were at decadal highs. One would have expected a high level of interest in these jobs. This has proven to largely not be the case due to the expected working conditions and has caused the authorities to take two approaches. One is to focus on attracting midstream and downstream companies to the area. Midstream includes aspects such as pipelines and associated pumping equipment. Downstream is defined largely by activities such as gas processing and conversion to higher value products. Shell has already announced an ethylene cracker for the region, although they have postponed it as of early 2015. The second approach to jobs is to attract regional headquarters of the operating companies. The township of Cranberry has had success with this angle.

Other Societal Issues

True adherence to the principles of sustainability would require that the community not only not be harmed but that it actually accrue some of the benefits from industrial enterprise. A case in point is the entire state of Alaska. Every resident receives a dividend check from the state representing a portion of the government revenue from the oil business. Families in Texas in effect receive a dividend: there are no state income taxes.

These examples will not necessarily find equivalence in other states because much of the revenue likely is from leasing of state land. Also, these payments are cash transfers, which are welcome by residents, of course. But it is arguably time to better integrate the benefits of science, technology, and industry within the nearby communities that have natural gas deposits. An example could be creating research stations relevant to the specific technologies used for extraction and transportation. Is there a way to create more positive spillovers for locals so the benefits are felt over time? These are complicated prospects, but fresh thinking on ways to do this may make natural gas extraction and fracturing much more palatable to communities.

Pennsylvania took a bold step. All the evidence points to them waltzing into prosperity provided the new rules are complied with. This is facilitated to some degree by larger oil companies buying out the smaller outfits. Larger companies have more to lose by missteps. Others will learn from this, but Ben Franklin's state is likely to capture the first-mover advantage.

CHAPTER 21

Will Cheap Natural Gas Hurt Renewables?

"The answer, my friend, is blowin' in the wind"

—From "Blowin' in the Wind" by Bob Dylan

Yes, cheap natural gas will hurt the rate of growth of renewable energy. There is no way to sugarcoat this unfortunate outcome. But much can be done to ameliorate the effect, and most of that lies in the policy arena.

Ironically, the celebrated successes of NGOs, primarily the Sierra Club, in shutting down coal plants and halting the building of new ones may hurt the cause of renewable energy. In the time frame required, in many instances wind and solar options will not meet base load grid parity. This is generally defined as parity with the cost of base load electricity production in that area. So, absent strict policy measures, which will be very hard to come by, natural gas will be the fuel of choice. Natural gas plants will not simply be mothballed when wind power reaches economic parity. The early demise of coal plants will lead to natural gas-fueled electric power plants with greater installed capacity.

In recognition of this, the Sierra Club had initially taken a position of natural gas as a bridge fuel until renewable sources became more economically viable. Hand in hand was the insistence on environmentally responsible production. In early 2012, Sierra Club executive director Michael Brune seemingly reversed that position, citing the environmental risks posed by shale gas production. He said, "As we phase out coal, we need to leapfrog over gas whenever possible in favor of truly clean energy" (Brune, 2012). While the "whenever possible" phrasing leaves some wiggle room, this is a definite movement away from the Club's original position.

In the view of this author, a life member of the Sierra Club, the solution is the responsible production of shale gas as bridge fuel. Clean coal is not an oxymoron, but it is expensive. At the same time, the move to renewable energy must be accelerated. In another odd twist, natural gas may be needed at first as a load leveler, even for wind and solar production, due to their diurnal cycles. Someday we hope to develop effective storage mechanisms. But someday can be a while, and we must be realistic regarding the need for bridging mechanisms. The 2012 Department of Energy budget included funding for an Energy Innovation Hub researching energy storage which was awarded to a group led by Argonne National Laboratory. This is good policy.

The price of natural gas in North America is roughly one-fourth that of $100-a-barrel oil on the basis of energy content. For this computation, I consider a more normal $4 per MM Btu, not the $2.50 in March 2012. In early 2015, the price of oil dropped to the range $50 to $65 per barrel, but gas also dropped, to $2.50 per MM Btu. Ultimately it is the low gas price that will hurt renewables, most of them being in the electricity sector. In Europe the price is higher but still a factor of 2 to 3 cheaper than oil. The discussion of whether renewables will be hurt falls in two distinct realms: electricity generation and transport fuel production. Accordingly, I consider these separately because the impacts will be distinctively different.

But first, let us examine the very premise of cheap natural gas. Until the shale gale swept us up, natural gas prices fluctuated considerably. For the last 13 years, the Henry Hub price was as low as $2 per MM Btu and as high as $13 (Figure 17). The Henry Hub price is the price of gas in Erath, Louisiana, and is used as the price for trading on the New York Mercantile Exchange. It is a rough proxy for US natural gas price.

Figure 17. The Henry Hub price per million Btus, 1999 to 2012

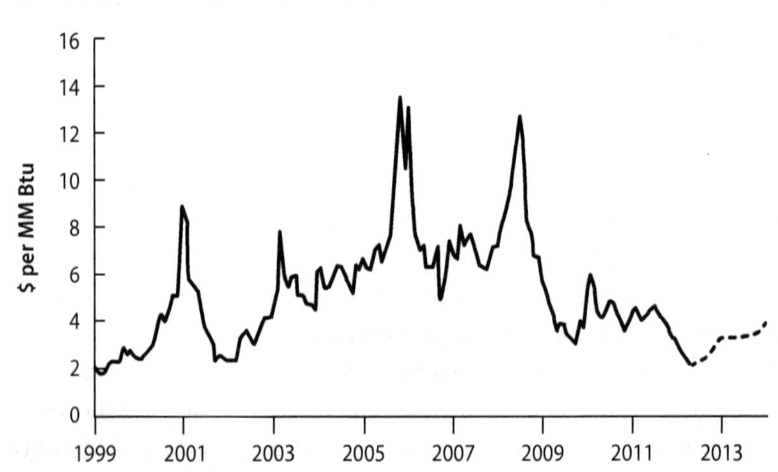

Source: US Energy Information Administration, May 9, 2012

The inherent uncertainty in gas price caused coal and nuclear to remain as viable options for electricity generation. Entire chemical industries moved abroad to regions of predictably cheap gas.

Gas from shale at first was costly. After the kinks got worked out and further advances were made, it could be produced profitably at costs lower than many conventional gas operations. Meanwhile, the sheer volume kept the price down. At the prices today the most profitable operations are those with higher proportions of natural gas liquids in association, because the value of this component is pegged to the price of oil and benefits from the high price of oil. The dry portion of the gas therefore continues to be produced, even at low prices.

Inevitably, consistently low prices will create demand and eventually prices will rise. In the face of this I offer the view that shale gas production will keep gas prices moderate. This is largely due to shale gas wells being on land and shallow by industry standards. These wells can be in production in 30 to 60 days after commencement. This short duration effectively keeps a lid on the price. If the three-month strip (the commodity price three months into the future) is seen as going up, new wells can be in production well within three months. This sort of certitude will also discourage speculative investment in the commodity.

Effect on Electricity Production

The floor price will get set by the conversion from coal to gas for electricity. Forty percent of coal plants not expected to meet the latest EPA standards on mercury and NOx are over 50 years old; these fully depreciated plants will not be refurbished. The only options are new coal, nuclear, and natural gas plants. New coal is disadvantaged on price alone until natural gas reaches a price of $8 per MM Btu. Recently in early 2015 that price has been under $2.50. So, with the ceiling mentioned in chapter 3, coal is not the economic choice. Nuclear has suffered a blow due to the Fukushima Daiichi disaster. So, natural gas will be the fuel of choice. Eventually, the shift to gas could cause the price to rise, but the lid will still be around $8. A wild card on price is the possibility of strong liquefied natural gas (LNG) export demand. This is not far-fetched, especially to Japan, where natural gas prices are about triple those in North America. The total added cost of liquefying, transporting, and re-gassing natural gas would be up to about $5 per MM Btu. This puts the landed cost at well below the market price in Japan. However, such export would require federal permits. Russian belligerence in the Ukraine in early 2014 caused calls for LNG exports to Europe. Over the course of 2012 and 2013, three permits were issued, adding up to 4.5 bcf gas per day. More are expected, possibly up to 10 bcf per day. By early 2015 that number is up to 10.5 bcf per day.

Cheap natural gas will also cause a shift from oil to gas whenever possible. This additional demand will keep the price up in the medium term. So, let us assume a price of $8 as the ceiling price. At this price, electricity will be delivered at a little under 7 cents/kWh. This is the price that alternatives will have to meet on a direct economic basis.

This natural gas–derived electricity price is lower than the fully loaded price of energy from new nuclear plants, which will be over 10 cents. Currently wind delivers at 9 to 16 cents, depending on where it is. Offshore wind may be higher yet at this time. Wind also often suffers from the need to add transmission infrastructure. This is especially the case for offshore facilities. There is also the celebrated case of T. Boone Pickens' terminating a major land-based investment due to the absence of definite plans to add transmission lines.

Strictly from a techno-economic standpoint, wind still has an upside. Engineered solutions are likely to drop the price from current levels. But it continues to suffer from diurnality, and so it needs to be companion to another source or to storage mechanisms.

Effect on Transport Fuel

Transport fuel today is dominantly produced from oil. Oil prices can be expected to remain high, in part due to the burgeoning demand from India and China. This demand is a direct result of the sustained increase in per capita GDP in India and China; a strong correlation exists between this metric and per capita vehicle ownership. On the face of it, this prediction strongly favors gasoline and diesel substitutes derived from renewable sources.

While only a few today, such as Brazilian ethanol, have achieved parity on a cost basis, there is room for improvement. This is particularly the case for drop-in fuels from sugarcane or sugar beets. Butanol is an example of a liquid that is a complete gasoline substitute in any proportion, unlike ethanol, which is more corrosive and water absorbent, and suffers a calorific penalty—that is, it has a third fewer calories than gasoline and commensurately worse fuel economy.

However, one does need about 8 percent ethanol in gasoline simply to act as an oxygenate, which makes the combustion of gasoline more complete. The chemical MTBE was previously used for this purpose, but has since been outlawed for environmental reasons involving pollution of groundwater.

Currently the most promising avenue for drop-in fuels is production of alkanes from sugar. The candidate raw materials are sugarcane, sugar beet, and sweet sorghum. Alkanes are straight-chain compounds with the formula C_nH_{2n+2}. Conventional oil-derived fuels have this formula as well. The number n is about 7 to 9 for gasoline and about 12 to 18 for diesel and a bit lower for jet fuel. So, alkanes with the right number are for all practical purposes direct drop-ins for these conventional fuels.

Herein lies the attraction. Also, being tailored, often through genetic engineering, the composition will be predictably uniform. This is not the case for the input to refineries from a variety of crude oil sources. In fact oil refineries today are forced to be very picky about the mix of crude they will accept. Seed-based oils also suffer from this variability.

No small wonder, therefore, that many of the leading players in the drop-in biofuels space are supported by major oil companies such as ExxonMobil, Shell, and Total—all heavy hitters.

Cheap natural gas can affect transport fuel from renewables in two ways. One is the direct use of methane as fuel for combustion in the engine (see chapter 17); the bottom line is that methane is unlikely to be a material factor for passenger vehicles until there are material advances in storage of CNG. The other is the production of transport fuel from conversion of natural gas, known as gas-to-liquids (GTL; see chapter 16); the point of discussion here is that cheap natural gas may be converted to transport fuel for a relatively low cost. If that cost is low enough it could disadvantage the schemes described above.

However, the proven technology in this arena, Fischer-Tropsch synthesis, is still expensive. Further research in catalysis will reduce its cost. Right now no compelling evidence suggests that alkanes from sugar will be particularly disadvantaged with respect to GTL, even from cheap methane. Finally, methanol from methane is a wild card, as discussed in chapter 19. Another wild card is methanol from biomass. Biomass-derived methane is likely to achieve renewable fuel status, and that will help.

In the end, the true yardstick will be comparison with the price of imported oil, not with other oil alternatives. Gas-based oil substitutes could coexist with renewable liquid fuels.

Policy Matters

Without a price on carbon, the carbon-free alternatives wind, nuclear, and solar are seriously disadvantaged. Taxes are anathema to Congress. Cap and trade has not worked particularly well in Europe, in part due to the uncertainty, which effectively increased the discount rate on investment. Also, any cap and trade conceived by Congress will undoubtedly have numerous exclusions and grandfathering. The province of Alberta in Canada has an interesting model. They tax high-carbon-footprint heavy oil production over a certain volume. The rate is $15 per tonne ("Go Figure," 2007). The money is placed in a special fund expressly for the purpose of addressing environmental issues associated with oil and gas. Such directed use of tax proceeds is more palatable. Conceivably, the fund could subsidize renewables for a period of time.

Finally, one could resort to the current method of imposing a renewable portfolio standard, which essentially mandates that a proportion of delivered power be from renewable sources. This in effect is a tax on the consuming public because the renewable energy costs more. The solar subsidy in Germany is passed on directly to the consumer as well. But that is largely possible due to the considerable influence of the Green Party. Short of taxing conventional oil and gas, consideration could be given to decreasing the incentives and redirecting those funds. But policy, if not based on an understanding of market forces, can have unintended consequences. An example of that is the "Flex-Fuel Fairy Tale" (chapter 25).

Conclusion

Cheap natural gas will place every other source of electricity production, including renewable energy, at a disadvantage for the short to medium term. Reliance on market forces alone will slow the introduction of renewable energy. Policy mechanisms are needed to level the playing field, at least from the standpoint of carbon neutrality. The most equitable methods may be a US analog to the method used in Alberta. By all accounts that policy is embraced by the public and industry alike. If the cheap-gas-enabled methanol displacement of gasoline and diesel becomes a major factor, even biofuels could be negatively impacted. On the plus side, methanol could be synthesized from biomass.

CHAPTER 22

Kicking Shale Into the Eyes of the Russian Bear

"You don't tug on Superman's cape"

—From "You Don't Mess Around with Jim" by Jim Croce

On January 7, 2009, Russia shut off the natural gas ("Europeans Shiver," 2009) flowing through the main European pipeline in the Ukraine. This was a particularly cold winter and 20 European countries encountered serious natural gas shortfalls. Discussed below are the reasons given by all of the players. But the principal point was, and continues to be, that Russia can use natural gas supplies as a weapon to achieve political objectives. In late 2008 Russia threatened to form a gas-based OPEC (dubbed OGEC) with Iran and Qatar with the express intent of manipulating world gas prices. Has shale gas dampened their ardor? Recent events in 2014 present mixed evidence on this point. The Ukrainian overthrow of its Russia-aligned leader caused Russia to take limited military action in Crimea. But as could have been predicted, they are using natural gas supply as a weapon by raising prices (reducing discounts) in the Ukraine. European response to the Russian aggression is muted compared to that of the US. In large measure this is because of the gas supply sabre that the Russians are brandishing openly.

Unilateral fuel cutoff as an instrument of political will would be essentially not possible with oil. Oil is more fungible, and alternative supplies can be brought to bear if a major supplier falters, deliberately or otherwise. It may cost more but you could get it. This is why Russia is unlikely to resort to any threats with respect to their coal and oil supply to Europe.

Natural gas is a regional commodity. Bulk transport across land can only be done through pipelines, and these are expensive and have long lead times. Transport across the ocean is feasible only if the gas is liquefied. (For shorter distances there are exceptions, where gas pipelines cross bodies of water, such as the North Sea.) The product is known as liquefied natural gas (LNG). This process entails cooling the gas to -161°C into a liquid that is 600 times as dense as free gas. This is then transported at close to atmospheric pressure. The low temperatures are maintained by auto-refrigeration through allowing small amounts to boil off, which causes chilling of the remaining liquid. An everyday analog is cooling of our skin by a fan or a breeze causing evaporation of our perspiration.

While LNG is a viable alternative to a domestic gas supply, it can only be delivered to a port location, and in fact only one with a re-gas terminal. The high capital cost of such terminals is unlikely to justify a capability merely to be available for upset conditions. So, as a practical matter, withholding of a domestic source is a powerful weapon, LNG alternatives notwithstanding. Also, LNG is more costly. Typically the added cost over the price of the gaseous version is about $3 to $4 per MM Btu. (Transport distance is the determinant of where you are in that price range.) As a frame of reference, that is roughly the price of natural gas in the US today. So LNG would essentially double that. This is why cheap shale gas in North America has rendered imported LNG passé.

The sheer distance between producer and user is the reason natural gas prices are so variable across the world. The price in Europe is about double that in the US, and in Japan the price is about triple and even more, as seen in Figure 18. Although I am not presenting a new figure, the numbers in early 2014 are very similar to those shown for 2012. The latest figures are likely impacted by the Fukushima Daiichi disaster–driven shift to natural gas for power. With the Japanese government's steadfastly negative position with respect to future nuclear plants, the price could remain high. In general, the prices are high in part because costly LNG is the marginal cubic foot, and so sets the price. "Marginal cubic foot" is industry terminology to mean the last tranche of gas added to serve demand. Since it is essential, the high price gets paid. This inevitably increases the price from all sources because the domestic producers can charge that figure. The exception is countries such as India and Argentina, where government-controlled prices artificially allow high LNG prices while prices for domestic production are kept low. The LNG business is built on long-term contracts because the capital cost is in the neighborhood of $4 billion. Nevertheless, there is considerable arbitrage; tankers will sometimes reroute in open waters to new destinations offering higher prices.

Figure 18. Trends in natural gas spot prices at major global markets

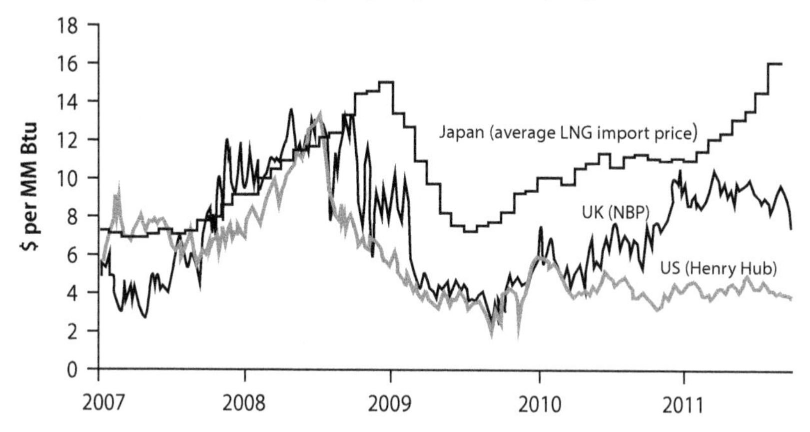

Source: Energy Information Administration, September 30, 2011.

Russian Use of Gas as a Weapon

Unlike in the Soviet era, Russia can no longer impose its political will through threatened military action. However, Russian gas is a significant natural gas source for most European countries. It is the dominant source for nine countries, including Greece, Finland, Hungary, and the Czech Republic. The recent imbroglio in the Ukraine has caused a number of these countries to seek alternatives. While this will take time, the movement is certainly under way. This monopoly allows unilateral action against any one of the countries. Action against too many would result in loss of needed revenue. (The Arab oil embargo in 1973 had a profound and lasting effect on the price of oil, aside from the short-term privation. But the original political objective was not realized, that of causing a significant shift in support away from Israel. Interestingly, though, the lasting price escalation that was a direct result of the embargo swelled producing country coffers. This allowed financing of politically motivated actions in other countries, including the funding of Islamic schools known as *madrasas* in Indonesia and other countries. These are believed by some to be linked to militancy.)

In an odd twist, the embargo-driven sustained higher oil prices opened up exploration in promising but costly areas such as ultra-deep water and the Arctic, thus reducing dependency on OPEC. Since then Norway and Brazil have become important players, on the backs of deepwater development.

The Russian action in 2009 was allegedly driven by a dispute with the Ukraine with respect to poaching on the gas line. While there may have been truth to this, most believe the action was intended to injure the Ukrainian Orange Revolution, which was seen by Russian President Dmitry Medvedev as not commensurate with Russian interests. The temporal connection with the gas cutoff action strongly implies causality. In many ways this act was more effective than would have been a military one. It also undoubtedly sent a message to other European states. This message was clearly heard. Germany in particular has not agreed with the US on the severity of economic sanctions in response to the Russian occupation of Crimea in 2014. Even Western Europe was affected in 2009, with southern Germany losing about 60 percent of its imported gas.

Shale Gas Could Change That

As discussed in a chapter 1, the mechanism by which shale gas accumulates makes it likely to be ubiquitous. So the likelihood of substantial deposits in Europe is high. Initial estimates by the EIA show large deposits in Poland and France, with smaller amounts elsewhere, including the UK and the Ukraine. Poland is actively exploring and the UK is following suit. France currently has a moratorium on fracturing, but is also not as much in strategic need due to low dependency on coal-based power. US efforts to produce gas with a minimal environmental impact will be important for widespread exploitation in Europe. Poland is certainly resolute on the matter. But their experience to date has not been very promising. Other countries such as Algeria, Libya, and Tunisia appear to have promise, and there is talk of a pipeline to supply the Ukraine. In late 2013 the UK government encouraged development, to the consternation of many. Furthermore, as in the US, as exploration proceeds, the resource estimates are bound to increase. All new hydrocarbon resource plays follow that pattern.

Gazprom, the mammoth Russian company operating gas assets, has publicly expressed concern regarding the effect of shale gas on future pricing. The fact that Russia too will have large deposits is irrelevant. Further increase in their resource base is interesting, but not a factor in the concern regarding domestic sources in client countries. To date Russian exploration of shale gas assets has been sporadic. They have used US service companies, which complicates US economic sanctions. Their shale oil assets are huge, estimated by the EIA to be the most of any country.

An interesting development is that US shale gas is increasingly being exported as LNG. Until European deposits are developed, US-sourced LNG could be a factor in offsetting Russian supply. If US prices remain low, as is expected, the cost of delivered LNG in Europe could profitably be at under $9 per MM Btu for some years; it is closer to $8 today in mid-2014. Furthermore, major LNG developments in Qatar and elsewhere which had been destined to supply the US will now find Europe a ready buyer. From a Russian standpoint this will not be a pricing concern, but certainly the gas as weapon argument is affected: a steady stream of LNG to Europe from new LNG plants would cut into the Russian gas market and certainly offer alternatives in case of threats to cut off supply. However, LNG takes years to be brought on stream. New capacity is not a credible response to such threats. Strictly from an economic perspective, the best sources for North American LNG are Alaska and British Columbia, and the most logical target customer is Japan because of location and price. That notwithstanding, the principal permits given to date in mid-2014 are for three Gulf Coast LNG importers turned exporters, adding up to about 5.5 bcf per day of consumption. My guess is that permits of up to 15 bcf per day will eventually be given. By coincidence that is exactly how much Europe imports from Russia.

OGEC Is Dead

Sixty percent of the conventional gas reserves reside in Russia, Iran, and Qatar. Operating costs are very low, especially in Iran and Qatar. In late 2008 the three announced intent to form a gas-based OPEC, which was dubbed OGEC. (Note: The *P* in OPEC is for Petroleum, and by definition, albeit not by common usage, gas is included in the term *petroleum*, so the acronym OPEC could have applied to gas as well in theory, but for the existing different cast of characters that would not have made sense.) Alexey Miller, chairman of Russia's Gazprom, said they were forming a "big gas troika." He also predicted an end to the era of cheap hydrocarbons, thus signaling the intent of the gas cartel to raise prices and keep them high. OPEC accomplishes this despite supplying only about a third of the world's oil.

The troika would likely have been pretty effective, in part because Russian markets are Europe and China over land, and Iran and Qatar are much more LNG-dependent. So, unlike current OPEC members who compete in the same markets, at least the senior partner, Russia, would be essentially noncompetitive with the other two, except for LNG relief valves for Russian force majeure, contrived or otherwise.

Shale gas over time will kill attempts at OGEC. China is expected to have even more shale gas resource than the US and will exploit it quickly. China National Offshore Oil Corporation (CNOOC) has already taken ownership positions in two US shale gas development projects and in the first large one in the UK. (There is little doubt that part of the intent is to transfer technology to China deposits.) Shell is partnering with Chinese entities to develop resources in the southern provinces. Argentina, the second biggest resource holder according to the EIA, will be a significant supplier over time, as I discuss in chapter 28.

European shale gas will certainly be a factor. There is reason to believe most of the countries currently importing LNG, including India, have shale gas opportunities. Finally, there is the reality of the US as an LNG export player; how much of a player is all that remains to be found out. All of this adds up to a world with a lot of gas in consuming countries and more options. When consumers have options, cartels are ineffective. Gas has always been harder to manipulate than oil. Transportation needs can only be met by oil-derived products. Gas on the other hand can be replaced by coal, wind, and solar for power. OGEC can be pronounced dead on arrival, and we have shale gas to thank for that.

CHAPTER 23

Shale Gas and US National Security

"Oh peace train take this country"

—From "Peace Train" by Cat Stevens

The biggest winner from a prohibition of fracturing and hence shale gas production would be Russia, closely followed by Iran. The ability of Russia to impose its political will upon the world, and Europe in particular, will be significantly enhanced. Absent shale gas, Russian gas will increasingly enjoy a monopoly in many European countries. Gas supply throttling could be used as a weapon of political will, as it was in 2009 and as is very much in evidence in 2014 with the Ukrainian crisis. Although the target was the Ukraine in 2009, the Slovak Republic suffered enormous collateral damage, estimated by some to be 100 million Euros per day for the 10-day period. This caused gas-deprivation-related recession. While much of this effect is attributed to unpreparedness, gas supply as a weapon is clearly real. This time around, in 2014, the issue is complicated with the additional military action, so the financial effect on the Ukraine of just the gas suppression is not clear as yet.

In a scenario where gas was used as a weapon, Russia would also have an increased ability to side with countries with interests opposed to those of the US. Already in 2014 the European position on economic sanctions against Russia is clearly colored by the need to maintain natural gas and other trade with Russia. In particular this could apply to not joining in sanctions or other measures against Iran. Consequently, Iran would be an indirect beneficiary of this power.

Iran would benefit more directly from a world without shale gas. Prices for gas will rise, and almost as importantly, will be unstable. Internal squabbles have delayed bilateral deals with countries such as India and Pakistan. Sustained shale gas production in the US has closed this country as a destination for LNG, thus closing a window for Iranian LNG exports in general. The LNG from Qatar and elsewhere originally destined for the US will find other markets that Iran may have targeted. In fact today, with shale gas accounting for about 40 percent of US domestic production, LNG contracted to be delivered to the US is being delivered elsewhere, including India.

India is perennially short of hydrocarbons. Absent shale gas, and possibly even with it, India will rely heavily upon gas from countries such as Myanmar, Iran, and Qatar. The first two have the potential to be pipeline services, although each is beset with the issue of traversing nations not friendly to India. That leaves LNG from Iran and Qatar. Any pipeline would be very expensive, especially if it follows an ocean route, as would be the case for an Iranian pipeline avoiding the Pakistani land mass. This would mean long-term contracts. Shale gas–induced lower pricing would dramatically reduce the pricing on these contracts, with attendant beneficial impacts on the Indian economy. Today India is importing gas at about three times the US price.

Any Indian dependency on Iran will strain US relations, although the expected relaxation by the US of the economic embargo related to nuclear weapons in 2014, if it holds up, will likely remove that tension. Any hiccup in the relaxation of the embargo, together with a long-term Iranian supply of gas to India, could result in reduced US influence in the subcontinent. The US-Pakistan relationship is already in some strife over drone strikes and related matters, and the US can ill afford this additional complication in the subcontinent.

Reemergence of OGEC

The Organization of Gas Exporting Countries (OGEC), comprising Russia, Iran, and Qatar, will certainly have the ability to be resurrected if shale gas production is curtailed. Around 2009, Russia was very open about using the cartel as a means to higher gas prices. Higher export prices for gas would put even more revenue into their pockets and those of Iran. The world has already seen the mischief wrought by Middle Eastern oil money following the 1973 Arab oil embargo–driven sustained higher oil prices. It would however be specious to suggest that OGEC-driven gas price rises would have similar effects, because gas simply is not as fungible as oil. But gas is a much more basic commodity in that it affects many walks of life, whereas oil is predominantly used for transport.

If LNG becomes a significant fraction of gas consumption, keeping shipping lanes open will now assume importance for this commodity as well. Military costs will for the first time constitute an externality for gas. Much as is now the case with oil, conflict in the Middle East could start having an impact on US natural gas prices.

Energy Security

The International Energy Agency defines energy security as "uninterruptable physical availability of energy at a price which is affordable, while respecting environment concerns" (International Energy Agency, 2012). From the standpoint of national security, energy security is an important part. While a physical threat to the populace in the form of attack is a part of national security, a large aspect is protection of the economic way of life.

Examining first the issue of affordability, domestic sources do not equate to low cost, at least in the case of oil, because it is a world commodity. Some countries create low prices through subsidies, but these often have consequences. In India, where kerosene is a way of life for cooking and other purposes, affordability is created with a subsidy. However, this is at the distributor level and so is subject to diversion of subsidized fuel to profit-making enterprises. Even were this not the case, this setup creates a burden on the taxpayer. Subsidies or inducements to use less ought to be at the consumer level. On the latter point, some commercial offerings are subject to a tiered pricing system, but this is not the norm for household power. With electricity it ought to be simple to provide the first tranche of power at low prices, thus ensuring essential heating and cooling for the less fortunate.

Affordability applies to industry in a big way. As discussed earlier, entire industries left the US when gas prices were high and variable. Oil does not elicit that reaction as much because there tends to be a world price because the commodity is fungible. Certain states have been able to attract industry almost solely by the low power costs in the region. A subplot in that case is that in some instances the average cost is low due to a preponderance of older, fully depreciated, coal plants. With the passage of time these will need replacement and costs will rise. The move to renewable energy will also raise costs. These effects can be dampened if shale gas–enabled prices stay low for natural gas.

The usual broad argument for drawing a bright line between energy availability and national security is that of support of oil-rich countries with dubious leadership. Protecting our way of life involves ensuring supply, which implicitly supports activity potentially adverse to the national interest. One of the positives associated with shale gas is that it affords us our way of life without swelling the coffers of the likes of Russia and Iran with gas revenue to create mischief.

The Baker Institute Study

A study by the Baker Institute (Medlock et al., 2011) in Houston describes a detailed analysis of the effects of shale gas on national security. They constructed three scenarios, of which I will analyze just two for simplicity: the base case of full exploitation of shale gas and the extreme case of no further exploitation beyond that in Texas and Louisiana. In each case they calculated the decadal averages for gas price for the next three decades. In the base case they showed the average price of gas at $5.84 and $6.46, respectively, for the 2020s and 2030s. These numbers are in general accord with my projections in chapter 3. But my key point is that not only will averages remain moderate, but the excursions will not be large. Stability is almost as important as low prices because this certainty drives investment. The reason the chemical industry is returning to the US is as much the certainty as the current low gas prices.

The Baker Institute study is probably a bit optimistic regarding gas prices in the scenario without widespread shale gas exploitation. The averages may be right, but the excursions, which they do not appear to model, will almost certainly be large due to reliance on imported gas and perturbations due to upset conditions in the exporting countries.

The Military and Energy Security Nexus

The military is the largest consumer of energy in the US public sector, consuming 5 billion gallons of fuel in 2010. Access is not really the issue even in times of tight supply. But it is incumbent on the military to reduce its reliance on fuel while at the same time not sacrificing operational effectiveness. This applies to all forms of energy, not just fuel for transport, although that is the one with the greatest imperative.

Shale oil production, combined with natural gas displacing oil and fuel-efficiency programs, is reducing the import of oil. Not far-fetched is the possibility that by 2025 the US will source all its oil from North America. As this day approaches the reliance on Middle East oil essentially disappears. Some believe (Michaud et al., 2014) that this will allow the US to reduce the policing of shipping lanes such as the Strait of Hormuz. Ending the naval support is unlikely, but sharing with other beneficiaries is seen as possible. Michaud et al further suggest that this might allow the US to shift emphasis to the Pacific, even to the extent of a maritime security arrangement with China.

During the Iraq War, there was great deal of public angst regarding the price of fuel for the war effort, and many in the supply chain got blamed. The fact is that a captain in a forward emplacement is not worrying about the price when he or she needs fuel urgently. The monetary cost aside, the human cost of such delivery is substantial. In the Iraq and Afghanistan wars in 2007, an estimated 3,000 military and civilian support personnel were killed or wounded while transporting fuel or water. Reduction in fuel usage, substitution with more benign alternatives, local sourcing of energy and water—these all ought to be priority strategies for the military today.

Semi-permanent bases domestically and abroad could even invest in distributed fuel production. If natural gas were readily available, small-footprint production of a drop-in fuel would not be out of the question. Given this possibility, the military ought to fund such developments rather than massive coal- and gas-to-liquids projects. In any case, small-scale distributed power in the form of mini-nuclear (what would be more secure than a military base?), wind, and solar combined with a micro-grid could power entire bases off the grid. This would not only give a green feel, but also would render the base relatively impervious to weather- or sabotage-related grid outages. Certainly in forward locations, the solar option would apply.

Base vehicles are uniquely suited to fuel switchover. This is because infrastructure support for refueling is straightforward. Furthermore, in the example of CNG and LNG substitution of diesel and gasoline, where feasible the engines ought to be modified to take advantage of the high octane rating of methane, thereby delivering more power and distance for less fuel. The same goes for electric vehicles. Again, distributed electricity is easier and an electric vehicle delivers 60 percent more miles per unit of energy consumed. Not only will imported oil be substituted for, but less energy will be used. Only certain vehicles may be suited to electrification, but any gains would also have the virtue of symbolism.

Fresh water transport to front lines does not get much press but is a tractable objective for reduction. The shale gas industry will be learning to deal with low-cost water sourcing and treatment. These advances could be used to advantage by the military. Saltwater aquifers are fairly ubiquitous, and the shallower they are, the less salty. The Defense Department ought to consider sponsoring developments of small-footprint desalination plants, especially targeting the types of salt water anticipated in theatres of action. Every president in the last decade or so, no matter from which side of the aisle, has drawn that bright line connecting energy security and national security. President Bush, a champion of oil and a one-time owner of oil interests, famously complained about our "addiction to oil." President Obama has said,

"America's dependence on oil is one of the most serious threats that our nation has faced." That sounds like a national security statement. So, equating national security to energy security and thence to reduction of imported oil will not be disputed by many.

CHAPTER 24

Sustainable Development: A Double Bottom Line, Plus Afterthought

"For the times they are a-changin' "

—From "The Times They Are a-Changin' " by Bob Dylan

The definition of sustainable enterprises is the so-called triple bottom line, wherein economic, ecologic, and community benefit are all considered and balanced. Is that last leg of the stool given mere lip service, or is the energy production industry recognizing this element fully? And ought it to be?

The economic consideration is a given. Without that there is no profit, and absent profit, no enterprise. The ecologic or environmental piece is much in evidence today, and few new energy enterprises would dare ignore this element. The societal element is harder to define. One is tempted to think that this is strictly composed of negative impacts upon society, because that is where the rhetoric is directed. In some ways it suits detractors of the goal of sustainable enterprises to cast it in this light rather than a more generic one. So, for example, visual pollution is denigrated as a personal preference rather than as pollution in the classic sense.

The Reality of Visual Pollution

Perception is reality, the saying goes, and marketing folks well know that this is a powerful adage. One cannot bully people into feeling a certain way. Certainly not in commerce. But on an issue of alternative energy, some nudging, in the sense of Thaler and Sunstein, is in order. Richard Thaler and Cass Sunstein wrote a powerful essay, "Libertarian Paternalism," in the top economics journal *American Economic Review* (Thaler & Sunstein, 2003). Non-economists, such as I, must not be daunted by the staid prominence of the journal; this is an easy read. A further easier read, one that costs some money or trouble (going to the library) is their book *Nudge*. Basically they posit the notion that given free choice, people generally do not make the best decisions for themselves, even in an economic sense. People need to be given a nudge. My point is that just because folks feel a certain way about visual pollution does not mean they cannot be nudged to a different position.

One way to do that is to clarify the options. Until recently the Sierra Club was against coal, nuclear, and hydrocarbons in general. (Incidentally, coal is a hydrocarbon, but one challenged in hydrogen content, and many think of it as a different species, but it is not.) Last time I looked, that position was taken as tantamount to suggesting we switch off the lights. Wind and solar are great options. But they are still fledgling and incapable of base load service. In the interest of fairness, the Sierra Club until recently supported natural gas as a transitional fuel, to the consternation of much of the membership. Even more lately that position has softened to one of "leapfrog over gas whenever possible in favor of truly clean energy" (Brune, 2012). This is a laudable sentiment that will be tough to execute in time for all the forthcoming coal plant closures brought about in part by Sierra Club activism.

Duke professors, made famous by their paper (Osborn et al., 2011) connecting well water methane concentrations to shale gas production, suggested in an op-ed piece in *The Philadelphia Enquirer* (Jackson & Vengosh, 2011) that we eschew shale gas in favor of wind and solar. No matter that each of these has opposition as well. There are entire communities that will not permit a visible display of solar panels on homes. Wind power has long been opposed on visual lines. North Carolina, the home state of the aforementioned professors, has a law preventing wind farms on mountain sites, known as the Ridge Law. Many communities have strong opposition to offshore wind production in sight of land.

When one flies into Amsterdam Airport, one sees an abundance of wind farms in the water. Personally, I think they look like a flock of birds—but I am a techie, what do I know? Perhaps their acceptance is premised on the Dutch having had windmills as a way of life on farms. More likely is the explanation that it is a choice between that and Russian gas. In Holland that may not be the direct option, but in Greece, which is predominantly dependent on Russian gas, it would be. Southern Germany still has frigid memories of when the Russians capriciously shut down the pipeline through the Ukraine in the early days of January 2009 as a political move. The 2014 upheaval in the Ukraine and Russian response in June to cut off gas supplies ought to provide an impetus toward alternatives. Sadly the initial reaction has been solely to seek gas from other sources. In general, opposition to something should come hand in hand with a consideration of an alternative.

Societal Benefit

Fair and equitable economic benefit to the local and regional communities ought to be a goal of sustainable energy development. In Australia's Northern Territory, uranium mining has provided a dividend to the native Aborigine community, conjuring up the image of traditionally garbed locals riding on the beds of Toyota trucks. Every resident of Alaska gets an oil-related dividend of substance. But these are the exceptions.

One measure would be similar to that in Alaska. Royalties on production would in part be distributed to the county in question. At the very least, this would go to ameliorate some of the damage to infrastructure. In the case of shale gas drilling, the principal damage coming to mind is the deterioration of lightly constructed farm roads by heavy trucks carrying sand or water. This will be a problem regardless of how well the water-related concerns are handled. Beyond the issue of mitigation of damage, the community as a whole ought to benefit in some measure from the overall enterprise. The fortunate leasers of mineral rights should not be the only ones to benefit. That sort of inequity is a sure recipe for neighbor turning on neighbor, particularly when the have-not neighbor incurs some direct negative consequences of the activity.

An interesting thought on revenue generation for broader distribution is the adjustment of valuation of the properties generating revenue from the production. That property is now arguably more valuable on resale than before. (The only argument against it is that some will consider production equipment on the property as a negative.) The portion of the increased tax from this aspect could go into a special pot for the purposes discussed above.

In general, taxation related to shale gas production will be the subject of considerable debate. The producing camp will argue that taxation will diminish activity. But anecdotal evidence to date does not support this belief. West Virginia has been much more aggressive in this regard than Pennsylvania, and the drilling activity comparisons are in West Virginia's favor. These single-variable analyses are fraught. In this case, the higher activity in West Virginia may be in spite of the taxation, and driven by the higher quality of the fluids.

Separate from these arguments is the inherent societal benefit from affordable energy. In that sense, the very fact of cheap natural gas represents a benefit for all.

Technology Forks in the Road

Technology choice can often have a direct effect on the local populace. These forks in the technology road fall into two broad categories: benefitting the local environment and aiding the local economy. The first one is an easy choice if other things are about equal. An example of that is in fracturing operations associated with oil or gas production. As the industry became more skilled at drilling horizontally, the increasing reach of a given well allowed a new technology, known as pad drilling. This involves drilling and producing from up to 25 wells from a single location, known as a pad. The number of roads needed drops as does the areal extent of the effects of traffic. Also, this aggregation of wells allows for better supervision and oversight to minimize mistakes. Pad technology was developed in Colorado for the express purpose of minimizing road footprint. It now is even more important in farming communities such as in Pennsylvania.

Biofuels could face similar forks. The conventional approach would be to transport the biomass or crop great distances to giant chemical processing plants. The energy densities of these materials are inherently low. Consequently, the costs and logistics of transportation are high. Technologies are being developed to build more biomass processing plants close to farmland, to bring the mountain to Mohammad, as it were. These must be specialized to not incur the penalties of reduced scale. A few such are in early-stage development. This will not only reduce road transport, but also it would create local jobs, which in many instances are high-paying ones.

Distributed power is another example. Small 50- to 100-megawatt plants using biomass, wind or mini-nuclear, to name a few, could supply localities. In the limit they could eliminate the need for costly and unsightly transmission lines. At short distances, direct current (DC) would be a viable and preferred option to alternating current (AC). Edison would have smiled. For those not familiar with this arcane bit of technological history, this country faced a critical decision in the late years of the nineteenth century. The AC/DC choice was on the table for long-distance power transmission. Pitted were Westinghouse Electric and Thomas Edison. Edison lost that one.

Not far-fetched is the notion of developing technologies specifically fit for the purpose of benefitting host communities and nations. An example relevant to the latter would be the deliberate design forks taken to allow for manufacture in that country. One could accomplish this by selecting materials locally available. A design assembly could be modified to use low-grade steel as opposed to high-alloy steel that would have to be imported. Assemblies could be simplified to allow local assembly. In chapter 14, I discussed the distributed processing option for monetizing the ethane found in wet shale gas. That particular option is somewhat hamstrung by the immediate absence of a business model to execute. But these things are traversable, and there is little doubt that local jobs will be created.

In summation, the societal benefit component of energy alternatives need not be an afterthought. Many elements can be brought to bear with no adverse consequences to the economics of the enterprise. Also, the lasting value of being a good citizen cannot be underestimated. It's simply good business.

CHAPTER 25

Flex-Fuel Fairy Tale

"Dream on"

—From "Dream On" by Aerosmith (written by Steven Tyler)

The Utopian State, known the world over as the US, was in the throes of a dilemma. Much maligned for not doing enough to limit carbon dioxide emissions, it developed a plan that seemingly in one fell swoop tackled global warming associated with automobile emissions while at the same time reducing import of oil from nations, some of whom were deemed unfriendly, at least in the rhetoric of elections.

This solution was known as the 20/10 plan. The goal, to replace 20 percent of gasoline with ethanol in 10 years, was seen as visionary, if for no other reason as that 20/10 was about as good as one got with vision. However, even before vast quantities of alcohol had been consumed, a hangover of major proportions was in the making. Therein lies the tale.

The Utopian State, as befitted its name, was inclined to believe that the public would recognize a really good thing when they saw it. They especially believed in the maxim "If you build it, they will come," because said maxim was irresistibly derived from the powerful combination of Kevin Costner, the national sport, and mysticism.

So they built it, a complex web of subsidies to farmers, automobile companies, and refiners, and tariffs on imported ethanol, all designed to produce domestic ethanol to blend with gasoline, and vehicles that would run on the stuff. In a nod to perceived consumer preferences, they incentivized the auto companies to make flex-fuel cars, capable of using regular gasoline and also E85, a blend with 85 percent ethanol.

They even created demand for these cars by ordering their agencies to use them and mandating the use of the new fuel. Waivers to the mandate were given generously, no doubt in the Utopian belief that said waivers would not be sought if not merited. It seems that some of these agencies are seeing a net increase in gasoline usage (Kindy & Keating, 2008), a result contributing in no small measure to the aforementioned hangover.

At the core of Utopian belief is that folks will "do the right thing." So, purchasers of flex-fuel vehicles were expected to purchase E85, even from filling stations some distance away, ignoring the fuel consumption getting there and back. Then word filtered through that E85 delivered 28 percent fewer miles per gallon. In short, it was more expensive to use and harder to find. They started filling up with regular gasoline because the flex-fuel vehicle allowed that, and filling stations noted the drop in volume and stopped stocking E85.

This nightmare scenario was interrupted by the seemingly sudden realization that natural gas could be produced very cheaply from shale, a rock previously deemed too hard to produce from. US industry knew how to routinely convert natural gas to methanol. This type of alcohol could also be blended with gasoline to make an E85 analog dubbed M85. It was even worse on gas mileage than E85. But methanol from shale gas was so much cheaper than gasoline that the cost per mile driven was less. One had to refuel more often, but at least the public was given the choice. Choice was good but not compelling enough for filling stations to change their design.

The turning point came when Prof. Wunderbar from a prestigious Eastern university invented a small engine that led to a car running on M85 that delivered both fuel economy and the muscle of a larger engine. The design took advantage of the high octane number of methanol (117 versus 87 for regular gasoline), which allowed effectively high compression ratios, which in turn improved the efficiency of combustion. The result was elimination of the gas mileage penalty from using methanol, increased power for an engine of given size, and retention of the improved emissions associated with methanol usage. And all of this was achieved with a fuel that was consistently less than half the price of gasoline.

Auto makers vied with each other to retool and produce these cars without any federal incentive because the public actually wanted them. Fuel distributors rushed to install M85 pumps and realized that this was simply achieved by eliminating one grade of fuel. They came to the realization that all vehicles on the road today specify either 87 or 91 octane. A third grade (89) was not needed, and the third pump was now available with modification to dispense M85. The US government, not wanting to be left out of this, set policies to further these steps.

Shale gas development technology improved to where low-cost natural gas was assured for decades. This certainty with respect to methanol price staying low allowed massive investment in vehicles and methanol production and delivery infrastructure. All was well again.

And then they elected a president who resolved never again to set policy that was not market-based. The country united behind him on this, and it was never quite the same again. The country was henceforth known as the United States.

CHAPTER 26

Displacing Oil

"Killing me softly with his song "

—From "Killing Me Softly With His Song" by Roberta Flack (written by Charles Fox and Norman Gimbel)

Oil will be displaced but not entirely replaced in the foreseeable future, if ever. Oil dominates the transportation sector. This is enabled by engines that run best when employing an oil derivative such as gasoline, diesel, or jet fuel. Substitutes do not function as well because the engines have been optimized for those fuels over the last century. So, even if one were to make substitutes at a per-gallon cost comparable to the conventional fuel, they would suffer performance penalties. For one, the distribution infrastructure is not well suited to substitutes. In general we could refer to all of this as the *power of incumbency*.

Toppling powerful incumbents is rarely achieved in a short space of time and equally rarely by any one stratagem. Displacing oil in transport will also be incremental in part due to the existence of a massive legacy fleet optimized for it. But shale gas, which has recently burst into our collective consciousness, has the potential to be our "Arab Spring" in testing the supremacy of the oil monarchy. Few now doubt the staying power of shale gas to be an abundant low-cost hydrocarbon, provided completely feasible environmentally sustainable practices are followed. But one verse does not make a song. This discourse is about the confluence of innovations in technology and thought designed to remove oil from its monopoly and render it just another useful commodity (Luft & Korin, 2009).

Having predictably low-cost natural gas for decades is a concept so alien that investors and producers alike are still grappling with the implications. Support for this belief is now to be found in a few places, including the chapters of this book. Predictably low-cost gas over the useful lifetime of processing equipment will embolden the production of fuel substitutes with current known gas-to-liquids (GTL) processes. By definition, these produce fuels indistinguishable from the conventional versions except for being cleaner.

The truly exciting possibilities center on the production of fuel substitutes such as ethanol, methanol, and dimethyl ether (DME), together with the direct use of methane in gas or liquid form. Each of these delivers fewer miles to the gallon than the fuel it replaces. But each of these has much higher octane (or cetane, in the case of DME) ratings than the commensurate conventional fuel. This book provides a recipe for gradually chipping away at the dominance of oil. Ingredients include breakthrough innovations that challenge the orthodoxy in the production, distribution, and use of transport fuels. Key federal and state policy enablers are also important for rapid change.

The orthodoxy referred to here is the underlying dogma that dominates each portion of the value chain. This chain begins at the oil in the ground. The wells are drilled and the oil lifted out to the surface. It is then transported through a system of pipelines to refineries. These are massive chemical plants that process the oil into useful fuels such as gasoline, diesel, and jet fuel. Refineries tend to be concentrated in just a few geographical locales. In the US, the vast majority are on the coast of the Gulf of Mexico, although there are smaller pockets in the Midwest and the two coasts. So, by design, no matter where the oil is pulled out of the ground, it is transported to these locations for processing. The thinking likely was to place the refineries close to the consumption points and move the crude oil the needed distance. This made sense until the last few years. What changed was the nature of the oil.

Oil is not oil. The key point is that oil ranges from light oil, which has the consistency of honey in the summer, to extra heavy oil, which has the consistency of chilled molasses. (For those not accustomed to Southern cuisine, this stuff goes "glop" when poured out of the bottle.) The heavier liquid simply will not flow down pipelines because of its viscosity. To make it flow, expensive thinning chemicals known as diluents are required. Also, the more viscous a fluid, the higher the energy expended, and the attendant cost, in pumping it. All of a sudden the old paradigm does not make much sense. In an odd twist, in North America the heavier grade of oil travels much longer distances to refineries than does the light stuff. This sort of thing is the result of entrenched dogma. In chapter 14 I discussed this matter in context of the fact that pipeline infrastructure is inadequate to handle this new oil source. The Keystone XL pipeline, whose purpose in part is to serve the Bakken shale oil, has become a political football, its fate unresolved as of this writing. Even as oil is being displaced by alternatives, much can be done to make the original process more efficient. This is because of the key word *displaced*. Oil will not be replaced for a very long time, if ever.

The alternatives to gasoline are also conventionally produced in massive chemical plants. Much in the same way as for oil, natural gas is also transported great distances to these plants using a system of pipelines. Leakage from these pipelines, especially when they get old, is implicated in some models of global warming (Howarth et al., 2011) due to fugitive methane emissions. (Over a hundred-year span, methane is about 21 times more potent as a greenhouse gas than is carbon dioxide.)

Minimizing pipelines would have significant benefits when the producing zones are in areas with greater density of farming or homes, as is happening with shale gas in Pennsylvania, New York, and Ohio. Pipelines can be minimized by locating either big plants in the same state as the production or smaller plants distributed close to the wells. This latter is technically difficult, however, because most processes benefit from economies of scale. Simply put, bigger plants are more efficient. Challenging what in effect is a truism in chemical engineering practice used to be unthinkable. Not any longer.

In chapter 18, I described the progress being made toward a future of jobs all over the country, not just in a few concentrated pockets. A future with job creation in about half the time for conventional plants. A future with jet fuel being synthesized at each air base. The key enablers for this aggressive agenda are small-footprint processes for the production of fuel: Instead of custom builds at each location, small modules are mass manufactured and merely assembled in place. In essence, economies of scale are being replaced by economies of mass production and remote control systems.

The foregoing notwithstanding, the future will hold a mix of conventional and small-footprint methodologies. While the production method may be the difference maker in some instances, the more important concept is that of displacement of oil-based transportation fuel. The main driver will be economics. The 2013 forecast by the Energy Information Administration (EIA) calls for a steady increase in the price of oil for the next couple of decades (EIA, 2013, June 13). This, coupled with predicted low prices for natural gas, is the key incentive for natural gas–derived transport fuel. A secondary but significant driver is the regulation requiring lower tailpipe emissions. GTL-derived liquids are low in contaminants such as sulfur. Methanol has that advantage and additionally burns at a lower temperature, especially if one uses the stratagem of evaporative cooling, as described in chapter 27, "More Is Better." Some of the nitrogen in the air used for combustion is expelled from the tailpipe as nitrogen oxide (NOx), which is a pollutant. Lower combustion temperatures result in less production of NOx.

As mentioned earlier, all the substitutes have lower energy per unit volume. In other words, they deliver fewer miles to the gallon than gasoline does. In the following chapter I discuss the viability of a fleet of high-compression engines uniquely taking advantage of the high octane ratings of methanol, ethanol, and methane. Aside from aiding in the use of a more environmentally responsible fuel, cars with high-compression engines would simply be more efficient. Using less energy for the same gratification is a worthy goal in all walks of life. Any reasonable carbon emissions goal for 2050 will require that at least 40 percent of carbon mitigation come from simply using less.

Electricity producers have taken this message to heart, and efficiency improvement through the use of waste heat is common in coal, natural gas, and nuclear plants. All new plants have these features. The Energy Star program encourages more efficient devices such as air conditioners, home electronics, and appliances. Building insulation standards are in continuous improvement. In the transportation sector, improvements have been sought through federal targets on fuel efficiency. Much of this has been achieved with lighter materials. But the big prize of engine efficiency improvement through compression ratio increase has not been taken seriously. Systematic increases in compression ratio have not been seen for many decades. Even when flex-fuel vehicles were introduced to facilitate the use of corn-based ethanol, the high-octane rating of ethanol was not capitalized upon. A flex-fuel vehicle running on E85 (85 percent ethanol, balance gasoline) would easily have tolerated a near doubling of the compression ratios to the vicinity of 16:1.

I am not going to trivialize the difficulty of turning that massive ocean liner known as the automotive industry. The finely tuned assembly line manufacturing does not welcome change. The chassis is rarely interfered with. Seemingly new lines are built on old chassis. Back in my day, the Chevy Camaro and Pontiac Firebird shared a common chassis. Even the revolutionary electric car, the Tesla Roadster, is built on the design of the chassis of an existing Lotus sports car. Our proposal would require changes to the design of engine cylinders and operating controls. These are less daunting, and in fact Mazda voluntarily introduced an engine with a compression ratio of 14:1 to improve fuel efficiency. With some clever engineering, they can operate this on premium gasoline. A flex-fuel version would run like a top on E85 or M85, and in the case of the latter, for sure, would be a good deal cheaper.

But operating a fleet, no matter how efficient, that runs only on an alternative fuel is daunting. That is in fact being attempted by electric cars today, in particular the Nissan Leaf, which operates solely on electricity. In

2014 the jury is still out on consumer acceptance, and demand is proving weak for now. A car running on electricity alone is a marvelously simple device: no gear box, no transmission, and no differential if the design has a motor for each wheel. It will be simple to manufacture and maintain: no oil changes, no air filter changes, no catalytic converter malfunction lighting up the idiot lights on your dashboard. The biggest draw from an environmental standpoint is not even the zero tailpipe emissions. That is nice for sure. Particularly nice for the youth counselors at the neighborhood YMCA engulfed in the exhaust from idling engines of parents picking up tots. Ironically an electric car would simply not idle; it would switch off when stationary as do hybrids today. But the compelling value is in the fact that a pure electric vehicle is 60 percent more efficient than an internal combustion engine. This is on a fair-minded mine (or oil well)-to-wheel basis; the calculation is shown in chapter 17 for those inclined to be skeptical, as would I have been had I not run the numbers.

In portions of the world not served by an electricity grid, limited power is supplied using diesel or kerosene. In India this is largely subsidized so the apparent cost is not high. But the fully loaded cost to the nation is substantial, particularly because most of the oil is imported. In chapter 18 I discuss the feasibility of small-footprint synthesis of fuel using the raw material of convenience. This could include natural gas, methane from waste, and biomass. The target use would be for electricity generation, pumps for water wells, and cook stoves.

Direct use of natural gas in transportation has its supporters. For fleet vehicles such as buses and delivery trucks, the disadvantages associated with a low energy density are less in evidence. Cities such as Delhi and Kuala Lumpur have replaced diesel in public transport vehicles with compressed natural gas (CNG) with dramatic improvements in health outcomes due to the elimination of particulates in the air from diesel. For passenger vehicles, improvements in volumetric density will likely be needed for broad acceptance. The progress in this area is described in chapter 17, and the future appears promising.

In reading this chapter one could be left with the impression that this is an ode to natural gas. That is not the intent. I see cheap natural gas as an avenue for the displacement of gasoline and diesel with substitutes. For these substitutes to acquire currency, we need distribution systems. We also need acceptance by the consumer. Take the example of methanol as a gasoline replacement in a conventional automobile. Cheap natural gas will cause methanol to be cheaper per mile driven than gasoline down to a gasoline price at the pump under $2 per gallon. However, the range for a car that ordinarily gives 350 miles will be closer to 200 miles with M85 use. Some consumers will trade that inconvenience for lower cost, some will not. Of course, high-compression engines will narrow that gap. I among others believe that the public will opt for the lower-cost, higher-performance option and fill up more often.

Once gasoline and diesel substitutes are firmly entrenched, alternatives to natural gas will emerge as the raw materials for their synthesis. One such alternative is petroleum coke, a residue from the refining of heavy oil that has little use. Heavy oil is an increasingly large proportion of oil production. Given that oil is not going away any time soon, and the marginal barrel is getting heavier by the day (except for the shale oil anomaly in the US), the utilization of otherwise useless components is a good thing. There is also a minor irony in using a waste product of oil refining to displace the portion currently considered valuable.

Woody biomass would be a viable raw material as well, as would be any cellulosic waste such as corn stalks and bagasse (the fibrous residue after sugar is extracted from sugarcane). Other candidates are purpose-grown crops such as switchgrass. These materials are especially hard to convert to ethanol, although much research is ongoing in that area.

The compelling point is this: All these raw materials, including methane, are first processed to an intermediate gas known as synthesis gas or syngas. After that the further conversion to methanol, DME, or synthetic gasoline or diesel, uses syngas as a starting point no matter what the raw material used to make the syngas is. This means that if a new era is ushered in by cheap natural gas, the door is left open for other materials, including renewables, to capitalize on the opportunity. This then is our true future, a transport world in which oil is a useful but not strategic commodity, and one in which more environmentally benign raw materials play an important, and possibly dominant, role.

CHAPTER 27

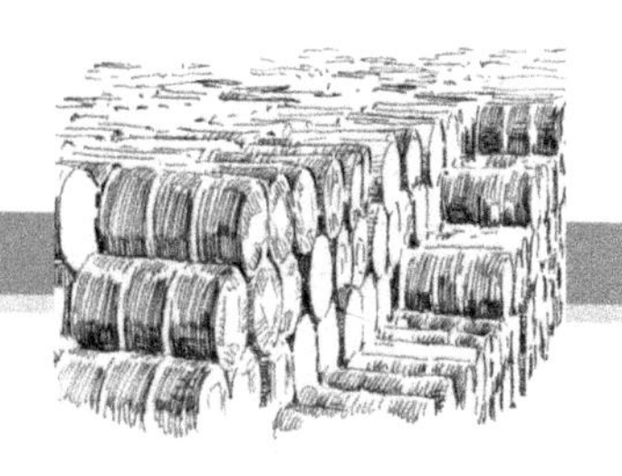

More Is Better

"My four-speed dual-quad positraction 409"

—From "409" by The Beach Boys (written by Gary Usher, Brian Wilson, and Mike Love)

Few would argue with the premise that more is better. A third more toothpaste in the tube, an extra shaving blade in the pack, two extra ounces of cereal in the box—these are merchandizing techniques that work. About the only one better is "free." Dan Ariely wrote an entire book based on this (Ariely, 2008). Incidentally, in your grocery store when you have a product marked "buy one, get one free," often you don't have to do just that. When you buy one item only, they are likely to give it to you at half price.

In the early years, automobiles ran on both ethanol and gasoline. The Ford Model T had an adjustable carburetor to allow this feature. This was in part a nod to the farming community, which could produce its own ethanol. Those early years of the 20th century saw considerable jockeying by oil interests. Ethanol was taxed for human consumption, but this tax was forgiven for industrial use. This was championed by President Theodore Roosevelt, who was openly critical of Standard Oil's domination. In 1920, prohibition introduced considerable pushback on even industrial use of ethanol because of the perceived risk of crossover into human consumption.

Higher-Compression Engines

Ultimately gasoline became the sole fuel in part due to the successful introduction of tetraethyl-lead (TEL) as an anti-knock agent in the 1920s. This allowed higher compression ratios, which improved fuel economy and engine torque, and hence consumer satisfaction. This idea—of increasing compression ratio for better performance—is the crux of this chapter. The box defines the term and explains the significance, but here I simply assert that the indicated improvements are a direct result of improved compression ratio.

Compression Ratio

The internal combustion engine running on gasoline has multiple cylinders that combust the fuel to produce energy for the wheels. Fuel mixed with air is injected into each cylinder. Then a piston closely fitting against the cylinder wall is pushed down onto the fuel mixture. The compression ratio (CR) is essentially a measure of how much the fuel-air mixture is compressed prior to ignition of the fuel. In gasoline engines, ignition is produced by a spark from a spark plug. The CR designed for an engine depends upon which fuel is expected to be used. Engines destined to run on regular gasoline these days have a CR of about 8:1 to 9:1. In earlier days and in some countries such as Brazil in the 1970s, CRs close to 7:1 were used. CRs in excess of 10 are deemed high compression.

At higher ratios, the pressure of the fuel-air mixture at the time of ignition is greater. This higher pressure results in more energy being delivered upon combustion and the piston being driven with more force. This translates into more energy supplied to the transmission, resulting in higher torque. Also, the overall efficiency of the engine is improved because more power is delivered for the same fuel use.

However, if regular gasoline with an octane rating of 87 is used in a high-compression engine, ignition may occur prior to reaching full pressure. This is known as "knocking" because of the audible sound created. The result is loss of power, plus undesirable carbon deposits. This is why high-compression engines need super gasoline, with octane rating of 93. Indy race cars have a CR of about 17:1. They use pure ethanol, which has an octane rating of about 113, or methanol, with an octane rating of 117.

The feasibility of blending ethanol with gasoline was known. Even the anti-knocking properties of ethanol were known. In fact, pure alcohol or a high-alcohol blend would have performed fine in a high-compression engine without TEL. But reportedly the oil interests that developed TEL successfully promulgated the view that TEL was the only solution. TEL was invented by General Motors, which joined with Standard Oil of New Jersey (later to become Esso/Exxon) to form Ethyl Corporation. The liquid containing TEL and some lead scavengers was named ethyl, presumably to not advertise the fact that it had lead. However, the gasoline was commonly referred to as leaded fuel. Much later, when TEL was replaced by MTBE, the fuel was referred to as "unleaded."

One could conjecture that the big driver for acceptance of TEL relative to alcohol was that only about 0.1 percent TEL was needed. For an equivalent performance by ethanol several percent of the fuel would be needed. Whatever the reasons, TEL-blended gasoline became the standard until over 40 years later. In the 1970s, the lead content in TEL was deemed environmentally unacceptable and a phase-out and replacement with ethanol was begun.

The phase-out of TEL induced another change in the engine. The fuel intake into the combustion cylinder and exhaust out of it was through valves. The sealing surface of the exhaust valves roughens and abrades, and TEL acted as a lubricant coating, presumably because of the lead produced, as noted in the box on the following page. Unleaded gasoline was expected to result in leaky valves. In recognition of this, the industry went to hardened steel for the valves and seats. Almost certainly the valve design detail was tweaked to ensure that the harder material still sealed effectively. (In general softer materials seal the best, which is why elastomers are commonly used for sealing. However, elastomers cannot be used in temperature extremes. For ultra-low-temperature sealing, especially of gases, the soft and relatively scarce metal indium is used. Certain specialized elastomers function at moderate temperatures of up to 350°C, especially if cooled down in some way. For very high temperatures, metal-on-metal sealing is used.)

This discourse on valve sealing and changes wrought by the move to unleaded gasoline is offered to demonstrate the ability of the automotive industry to respond to new requirements. Mass production methods are not very tolerant of change. But if it has to be, change in a few components is relatively easy to accept. This is the case with the design modifications required for flex-fuel vehicles, which again are primarily related to a change in materials, albeit in a less demanding situation than valving. Later in this chapter I discuss the desirability of modifications facilitating use of alternative fuels. These fuels are more complex, but the building blocks are already in the toolkit of automotive design engineers.

The toxicity of TEL centers on the fact that when combusted, the reaction product contains metallic lead. Prior to exhaust from the tailpipe, it more than likely would be in the oxide form or as a compound combined with the aforementioned lead scavenger additives. In either form, this heavy metal is toxic when ingested or inhaled. However, the driver for getting the lead out of gasoline was not the impact on health per se. Environmental health considerations resulted in the invention of the catalytic converter. This platinum-coated ceramic catalyst reacted with pollutants such as unburned carbon monoxide and nitric and nitrous oxides. Lead compounds deposited on the platinum and drastically reduced the useful life of the catalytic converter. The car industry, General Motors in particular, led the charge to eliminate lead from gasoline, commencing in 1974.

Lead and Lead Substitutes

Lead may have been eliminated from gasoline, but it is present in everyday life. All conventional car batteries have this element as the major ingredient. The lead-acid battery is still the most reliable source of power that can withstand the charge/discharge cycle over the useful lifetime of a few years. Car batteries are recycled to recover the lead. In fact, about half of all lead produced is from secondary sources. This insulates the public in large measure because the old battery is almost always returned to the store selling the new one. But in terms of the risk of emissions, the recycling process is not dissimilar to that used to extract lead from ores. This does not account for the environmental effects of open pit mining of the ore.

Lithium batteries are the latest to challenge the supremacy of lead-acid batteries. Cleaner, and currently more expensive, they have their own challenges. The grounding of the Boeing 787 Dreamliner in January 2013 was due to lithium battery packs catching on fire. This, too, will be solved. Scarcely anything of value replacing an incumbent ever comes free of baggage, at least in the early going.

Until the middle of the 20th century, one of the high-volume uses of lead was in paint of all sorts. Lead oxide could be made pure white or yellow or red. It also had the key desirable characteristic of paint: flows smoothly when brushed but becomes viscous and stays put when the brushing ceases. Ingestion by children and consequential health issues caused its gradual replacement by titanium dioxide. Today the majority of paint uses this extremely white substance as the base pigment. The best bond paper also uses it for whitening. In the US the material is derived from the ore ilmenite, which is a mixture of titanium oxide and iron oxide. The extraction method uses sulfuric acid to preferentially remove the iron as a sulfate, leaving behind the titanium oxide. Producers of the waste iron sulfate with residual acid, known as Copperas, have at various times been challenged for improper disposal of it. This is yet another example of benign substitutes having their own environmental issues.

TEL continues to be used in aviation fuel for piston engines. Modest amounts are sufficient for use in supercharged engines and currently no alternative is perceived to exist. Alcohol in gasoline has a tendency to absorb moisture. In certain situations the water can separate out. In the conditions of flight, the potential for ice formation in the lines is too risky. Also, alcohol would reduce the range, which for a plane could be prohibitive. In automotive applications the main issue would be corrosion of the fuel lines due to the moisture. This is avoided through the use of stainless steel tubing. All flex-fuel vehicles have this feature and additionally have more robust fluorinated elastomers in the sealing mechanisms.

Gasoline Substitutes Benefit From High Compression

The principal candidates for gasoline substitution are ethanol, methanol, and methane (natural gas with liquid components removed). Each of these has significantly less energy content per unit volume compared to gasoline. Adoption of these alternatives is hampered by this failing. A car running on E85 (85 percent ethanol and balance gasoline) will deliver roughly 28 percent fewer miles to the gallon. The comparable figure for M85 is about 42 percent reduction. The compressed natural gas (CNG) penalty needs to be expressed differently and is more fully discussed in another chapter. But the gist of it is that the tank has to be four times the size for the same range. The tank is thicker-walled as well and carries an additional weight penalty compared with a gasoline tank, which reduces range.

These fuels are consequently encumbered by the need to be less expensive per gallon. The terminology "dollars per gallon of gasoline equivalent" has entered the lexicon. Curiously, however, each of these gasoline wannabes has a valuable property that, if exploited, would overcome its shortcomings. That property is octane rating.

Octane Rating of Gasoline

As mentioned in the discussion of compression ratios earlier in this chapter, octane rating is the property of gasoline that defines the highest compression ratio that can be used. When crude oil is heated up, the constituents vaporize at different temperatures. These are then condensed separately to result in gasoline, diesel, kerosene, and so forth. Gasoline is in the lowest-temperature fraction. As distilled, it has an octane rating of about 70. That would safely run an engine with a compression ratio of only around 7:1, inadequate for modern cars. Refiners can process the "straight run" gasoline to produce some fraction of larger molecules, which have more energy density. The other approach is to add large molecules known as aromatics. These are so-named because the compounds have an aroma. Those with a pleasant aroma are fragrances. The aromatics added to gasoline are decidedly not fragrant. The class of compounds known as BTEX, which are environmental pollutants, fall into this category.

Some combination of processing and additives raises the gasoline octane rating to the value desired for regular gasoline, which is 87. This is suitable for the workhorse compression ratios of 8:1 to 9:1. Higher-compression engines need the higher grade of 91 octane. Some sporty cars require 93 octane. The added cost of the additives is passed on to the consumer.

The octane ratings of ethanol, methanol, and methane are, respectively, 113, 117, and 125. Reported numbers on these vary slightly, but these numbers are clearly much greater than regular gasoline at 87. When added to gasoline the effect is linear. This means that a 10 percent addition of ethanol to 87 octane gasoline will raise the value to 89.6. Does this mean that regular gasoline with 10 percent ethanol, common in most states, has an octane rating of mid-grade or plus? Almost certainly not. In recognition of this effect, the refiner is more than likely reducing his expensive octane-enhancing ingredients and relying on the ethanol in part to boost the octane rating to 87.

Flex-fuel vehicles, or FFVs, can operate on any mix of gasoline and alcohol up to 85 percent (that upper-end mix is designated E85). In Brazil, FFVs can operate with 100 percent alcohol. This measure was undertaken both in Brazil and in the US partly in response to the desire to reduce dependence on oil for transport fuel. The cost of conversion was low, on the order of $100 per vehicle. Since E85 was not commonly available in the early going, it made sense to impose the flexibility of being able to use straight gasoline. But since E85 delivers 28 percent fewer miles to the gallon, the consumer driven purely by economics would opt for the gasoline option. The result has been a paucity of E85 filling pumps.

Also, we have collectively missed a big bet: capitalizing on the high octane rating of E85, or M85 for that matter, which latter fuel would also function in an FFV, with some minor changes. These two fuels and methane (as CNG) would deliver very high efficiencies if run in high-compression engines. This improved efficiency would offset the lower energy content. Note that Indy race cars run on pure methanol. They have compression ratios (CRs) of around 17:1.

Flex-Fuel Vehicles with High Compression

Just in case this smacks of wishful thinking, consider a relatively recent commercial development. Mazda has introduced a line of cars with high compression ratios operating on conventional gasoline using an engine system that they have termed the SkyActiv engine. But Mazda's move appears to have nothing to do with enabling the gasoline alternatives. More below on how we can avail of that, no matter their intent. Mazda appears to have been shooting for higher engine efficiencies to meet the latest CAFE standard. Virtually all the other major companies were using electric vehicles and hybrids as the avenue to that goal. As I opine in the chapter relating to electric vehicles (chapter 17), battery costs are prohibitively high, and cost reduction to under $200 per kWh is needed for widespread consumer acceptance. Mazda appears to agree and has placed its eggs in the short term in the basket of efficiency through high compression. Since the design started years ago, somebody was reading excellent tea leaves.

What Mazda has done is to make the cylinder narrower and longer. But the critical innovation is to provide two injections of fuel during a single cycle. The second injection is right at the point when sensors detect temperatures reaching the point at which knocking (premature ignition) is imminent. When gasoline is injected into a hot chamber and evaporates, it sucks up heat to provide energy for the evaporation. An everyday example is the cooling effect on one's skin when aftershave lotion evaporates upon application.

The evaporative cooling drops the temperature enough to prevent premature ignition. This allows them to increase the CR to 12:1. Fancy exhaust system management allows them to go up another notch to 13:1. For 14:1 they need to use premium gasoline. They are offering this only in Europe. They believe American consumers will balk at the fuel premium, which is around 30 cents in most states. The Mazda3, offered with just the junior version of 12:1, is reported to improve mileage from 24/31 mpg city/highway to 29/39 mpg. There is some arm waving on manual versus automatic transmissions, improvements to the latter, and so on. But these are big improvements with no change in the gasoline.

So, what we have here is a commercially available car with a CR of 14:1 needing premium gasoline. It has substantially improved fuel mileage over the normal compression ratio car. If modified to be an FFV at a cost less than $100, this car would operate very comfortably indeed on E85 or M85. This would be an FFV with a realistic choice for the consumer: premium gasoline at 30 cents or no cost premium with E85 at the price of regular gasoline. With M85, the economic argument is even more compelling, as explained in chapter 19. Now for the punch line. If Mazda were to extend the technology to a CR of 16:1 or even 17:1, more than likely both E85 and M85 would function just fine, especially with the dual injection scheme taking advantage of evaporative cooling. This would deliver even better mileage at no additional fuel cost. Sure, it would not operate with gasoline, but neither do diesel vehicles. There is no reason to believe it would not be well suited to CNG.

We need to get out of the trap of finding gasoline substitutes for cars optimized for gasoline. An attractive menu would include the following. An FFV at CR of 14:1 or so, running on premium gasoline or E85, M85, or CNG, would be the moderate offering. An FFV with an ultra-high-compression engine operable only with E85, M85, or CNG would be another. CNG for passenger vehicles needs some advances, so that may be in the future. But these vehicles would be positioned for that. In the meantime, a fueling station would have regular and premium gasoline, together with a pump for M85 and/or E85. Of these last two, pure economics will drive the choice to M85 today. But if breakthroughs in cellulosic ethanol drive the cost down, the FFVs will be ready to accept that. Importantly, consumers will be empowered with fuel choice.

PART V

Next Steps

V. Next Steps

The oil price crash commencing in late 2014 could prove a blessing in disguise for the shale oil and gas industry. The US has already become the swing producer since early 2015. In other words, US production swings are deterministic of world oil price changes. A US oil surge could create serious winners and losers worldwide. In this climate of very low oil prices, displacement by alternative energy sources will prove more difficult and need policy support.

Energy policy is likely to undergo scrutiny and change in the US. Regulation affecting US export of US crude oil, the transport of oil by train, and induced seismicity related to waste water disposal are likely, at the state or federal level. Here I discuss the effects of various policy alternatives and identify areas needing further research. In addition, I discuss the effects of discoveries of deep sources of shale oil and gas in Argentina, currently a net importer of gas from Bolivia.

CHAPTER 28

A Dead Cow Kicks Up Its Heels

"Break on through to the other side"

—From "Break on Through (To the Other Side)" by The Doors

The Vaca Muerta in Argentina may well be the most promising shale oil and gas reservoir discovered to date. Strictly translated the name means "dead cow." But this particular dead cow is poised to make some noise, possibly a lot of noise. A hopeful President Cristina Kirchner refers to it as "Vaca Vive (Living Cow)." But the desolation of Patagonia is more in tune with the actual name. According to the US Energy Information Administration (EIA), shale in Argentina ranks third in the world for gas and fourth for oil. The Vaca Muerta alone is estimated by the EIA to hold recoverable quantities of 16.2 billion barrels of oil and 308 trillion cubic feet of natural gas good for 150 years at current Argentina national consumption rates. Earlier in 2014, the quality of EIA estimating took a bit of a reputational hit when it reduced the similar estimate for the US Monterey shale by 95 percent. In contrast, estimates in already-producing areas such as the Bakken continued to go up from 2007 to 2010. This propensity for increases in reserves estimates after development of a field has begun is normal because more information increases the certainty. This improvement in probability has a direct positive impact on the reserves figure.

The Vaca Muerta need not rely on the EIA estimates to prove its mettle. Exploratory and developmental drilling has resulted in no reported unpleasant surprises. The reservoirs are up to 1,500 feet thick, making them on average three times thicker than the Eagle Ford, with which it is rightly compared. The total organic content (TOC) runs a healthy 3 to 10 percent, rivaling the best performers in the US. The most interesting attribute may be the reservoir pressure. In that regard it is more like the Haynesville, which is deeper and hotter, and therefore more expensive. However, the Haynesville is also bone dry—quite unlike the Vaca Muerta, which, like the Eagle Ford and Utica, has the full spectrum of fluids ranging from dry gas to oil. In fact the Utica, discussed in chapter 4, may end up being the best analog, except for its lower TOCs. The depths are similar, and more importantly the expected reservoir pressures are comparable. Higher reservoir pressures mean higher production rates.

Mineralogy and Other Considerations

The mineralogy of a shale reservoir is an important characteristic for ease of recovery. The mineralogy of various shale reservoirs is depicted in Figure 19. All shale formations are some mixture of the constituents noted on each vertex of the triangle. Calcite and dolomite are carbonates of calcium and magnesium. Clay is an aluminosilicate, usually of sodium and potassium, and very ductile. Quartz is essentially sand (silica) in crystalline form. It and the carbonates are very brittle. For effective fracturing operations slight brittleness is desirable, but not too much. Plotted here are many of the known formations. The best ones are dominated by either silica or carbonate, but you can see that they are not close to either corner and there are none on the clay side. This is a technically difficult to explain format. Suffice to say near any apex, you have almost purely

Figure 19. Mineralogy of various productive shale formations

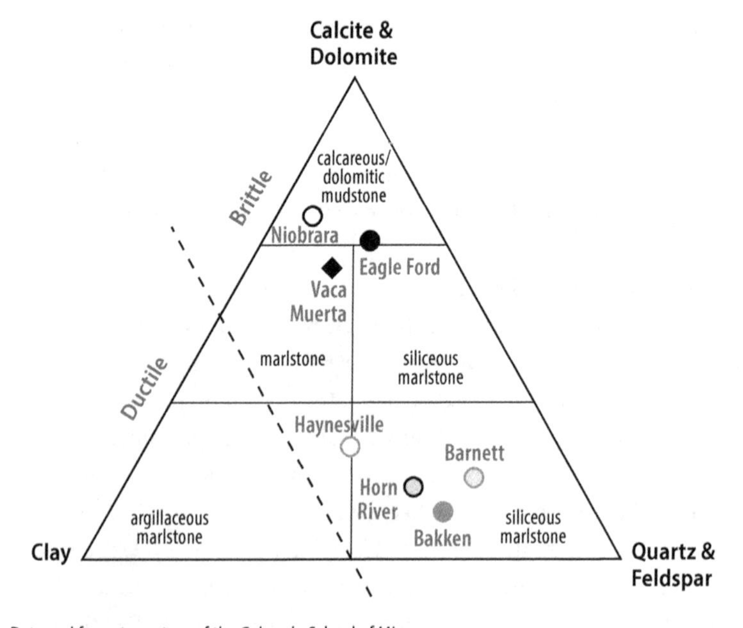

Source: Data and format courtesy of the Colorado School of Mines.

that substance. In the very middle of the triangle you would have a third of each constituent. The purpose is to compare with formations of known character.

Vaca Muerta is most like the Eagle Ford and Niobrara (which lies beneath much of the Great Plains of the US and Canada) in mineral character. Not shown are the results for the Utica, but it is also in the carbonate-dominated area. Production history for the Utica is still sparse, however, so the Eagle Ford experience may be the most valid in terms of predicting the comparative output of the Vaca Muerta. Results for Eagle Ford are shown in chapter 4, and the explosive growth is startling. Offsetting that in an Argentinian setting are the country's high inflation rates, estimated by some to be in the vicinity of 40 percent per year, although 30 percent may be a more generally accepted figure (Turner, 2014). Virtually all the drilling and production will be done by US-based service companies, and the pricing and availability of these services could be a factor. An interesting positive development in this regard is the growth of the Brazilian petroleum sector.

The Brazilian government requires that a percentage of the revenue be spent in-country on research and development. This has caused at least the two major service companies in the key area of hydraulic fracturing, Halliburton and Schlumberger, to set up significant R&D facilities in Rio de Janeiro. Manufacturing of service equipment in São Paulo has been ongoing and could be expanded. These two factors could play an important role, especially if the two countries choose to cooperate in this regard. The soccer rivalry will need to take a back seat, especially the perceived failures in the 2014 World Cup. In the case of Brazil it was total ignominy, an unceremonious exit before the semifinals. Argentina at least made it to the finals, but national hero Lionel Messi was not his World-Player-of-the-Year self, and Germany hoisted the trophy. The national psyche was assuaged by the fact that at least it had not been Brazil.

A Profile of YPF

YPF, the largest energy firm in Argentina partially owned by the Argentine government, is the dominant player in the Vaca Muerta. Together with US-based Chevron Corporation, it is developing the Loma Campana area. With an investment to date in mid-2014 of about $2 billion, YPF is producing over 25,000 barrels per day (bpd) of oil equivalent from 194 wells. The joint venture has committed to spend $16 billion over the next decade and a half. One would be tempted to make a linear estimate from the current performance and forecast production of 250,000 bpd. But if the US experience is a guide, and it ought to be, the cost per barrel of production will drop significantly. YPF's well costs at Loma Campana have already come down by over 30 percent from when development began.

YPF is led by Miguel Galuccio. What makes him particularly interesting is that prior to taking on the chief executive position, he was not an oilman in the classic sense, although he did work in oil and gas operations when younger. He came from running the Integrated Project Management division of Schlumberger and subsequently a variant of the same.

At one stage, I was in charge of the reservoir group of the Halliburton equivalent. The group conducted the sophisticated evaluation of a prospect in precisely the same way as did the oil companies. We then ran the drilling and production operations and were paid for performance, not work done by the day. Both Halliburton and Schlumberger had variants on the business models surrounding this principle. Both at various times incurred oil company ire, by being seen as competitors, and models had to be changed. A clear dividing line, at least at Halliburton, had always been to not compete for leases. The elephant in the room had clearly been the possible supplanting of the integrated oil and gas company's role as supplier of technology to national oil companies. Back in the day, the major national oil companies needed both capital and technology. In many parts of the world, after coffers were filled with production revenue, only the latter began to be needed. Argentina currently needs both.

At Schlumberger, Galuccio led a group that for all intents and purposes was an upstream oil company. His basic upbringing in a service company ought to allow him to be a judge of performance by service companies. This is of value in his role at YPF considering the fact that service companies perform virtually all the functions on a rig. Oilmen rising simply from the oil ranks would not have that insight from within, as it were.

YPF also has downstream operations, comprising three refineries of different complexity. It is very unlikely that Galuccio came with experience in that sector. But then, most chief executives of integrated oil companies (these are oil production companies that also own refineries) have come from one side or the other, usually the upstream side.

Galuccio brings a few other favorable attributes. He is an Argentine petroleum engineer who was born in the northern city of Paraná. His early work experience includes time spent with conventional production in Patagonia, home to Vaca Muerta. But curiously it is his experience abroad that will serve him in good stead. By all accounts he assimilates. This trait is not on every CEO search committee list, but it ought to be. For such people handling diversity becomes a way of life. While working in Malaysia he chose to acquire a working knowledge of Bahasa Malaysia (the country's official language). This effort may have played a part in the recent announcement of investment in Vaca Muerta by the Malaysian national oil company Petronas (Turner, 2011). An initial investment of $500 million in the La Amarga Chica oil field is likely to be followed by much more.

Too Much of a Good Thing

Sometimes it is better to be a follower than a leader. The folks in the front tend to get shot. Argentina is in a position to learn from both the successes and the failures of shale oil and gas production in the US. The implications relative to avoiding environmental pitfalls, with technology as well as regulation, are to be found in other chapters. Here I discuss how to deal with having too much of a good thing. This is not the same as the "resource curse," also known as the Dutch disease. "Too much of a good thing" is solely directed to the nature of the Vaca Muerta prospect. To be more precise, it relates to the fact that we expect the production of oil, gas, and natural gas liquids (NGLs) there. The oil will be light and sweet (low in sulfur). Most oil gets heavy and sour from bacterial degradation; the original "source" oil is therefore likely to be light and sweet. The NGLs will likely have a high proportion of ethane, as they do in the US. Both of these characteristics (the oil being sweet and the high proportion of ethane in the NGLs) have caught the US unprepared. Consequently, in 2014 US shale oil is selling at a substantial discount to the world market price for oil of this quality. Similarly, ethane is largely stranded, as discussed in chapter 14. Argentina is in a position to do two things. One is to construct simple refineries, as described in chapter 12, close to the oil production in the Neuquén and Mendoza provinces. This will create local jobs, add to the economy, and eliminate the need for crude oil pipelines. While finished goods pipelines may be necessary, the distilled products would in part be distributed locally.

The more interesting play is in the conversion of ethane to ethylene. This is a very straightforward process and can be done locally. The US was saddled with the fact that 33 of the 36 ethane crackers were located over a thousand miles away from much of the ethane production. The anomalous case of LyondellBasell is discussed in the box on the following page. Starting from scratch allows Argentina to place these crackers in the Neuquén province in close proximity to wet gas production. The country could go from being an importer of ethylene and derivatives to a net exporter. For a country saddled with a poor foreign exchange balance, this would help level it out. I have opined elsewhere (Rao, 2014) that monetization of natural gas components is particularly advantaged by vertical integration on a project basis. YPF, while in the refining business, would do well to acquire technology unique to ethane cracking, and associated capital, through such vertical integration. The principal tenet of the advantage of vertical integration is that when the producer and converter are partners, the raw material acquisition does not require complicated, long-term supply agreements and the producer can also benefit from the high price for the finished good. Ethane priced at $4 per MM Btu (the price in the US in mid-2014; see Figure 20 for trend) could be cracked for a cost of about $600 less per tonne than ethylene produced conventionally from naphtha derived from $100-per-barrel oil.

Figure 20. Ethane price trend compared to oil price (WTI), 2008 to 2015

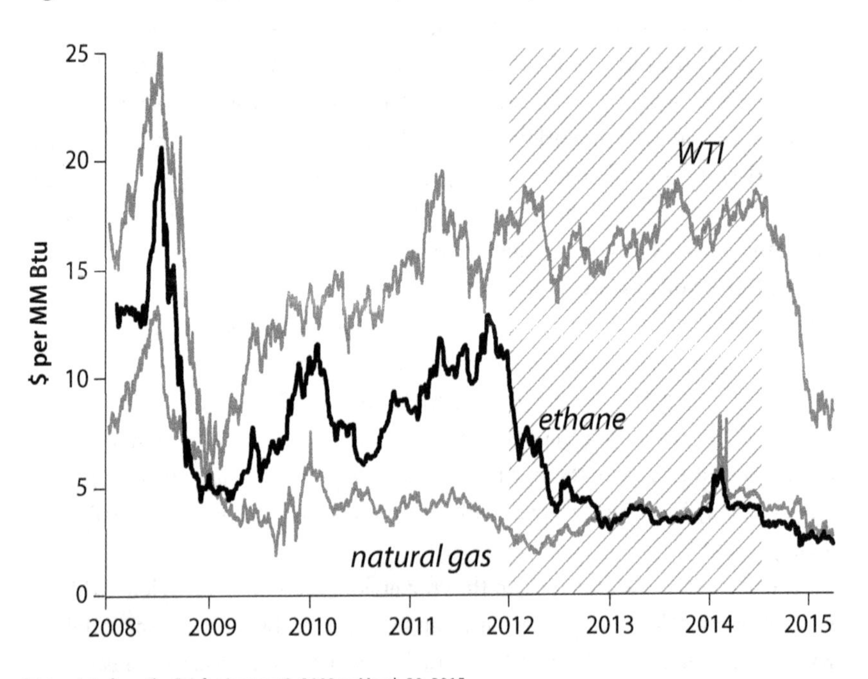

Pricing data from the EIA for January 2, 2008 to March 30, 2015.
WTI refers to West Texas Intermediate, a benchmark for crude oil pricing.
Source: Plot used by permission of David Bem and Mark Jones, The Dow Chemical Company.

The Curious Case of LyondellBasell

A recent story in *Forbes* (Vardi, 2014) tells the tale of the rise, fall, and rise again of LyondellBasell, a Houston-based chemical manufacturer. At first blush the story is a gripping tale of savvy investing, risk taking, and two opposite bets by equally shrewd investors, both of whom made large profits. But the true message is really none of that. It is entirely the plus side of the ethane dilemma discussed in chapter 14. It is also a lesson in the possibility that behind every problem lurks an opportunity.

The tale unfolded in 2007 when Basell, a European chemical company, bought struggling Lyondell Chemical Company. By 2009 the newly named LyondellBasell sought Chapter 11 bankruptcy. Apollo Global Management saw an opportunity and essentially purchased the company for about $2 billion. Meanwhile, Leonard Blavatnik, one of the principals in Basell, also bought into positions in the restructured company. The company took off, and Apollo cashed in about $12 billion in 2013. Not a bad profit in five years. Blavatnik, with a similar proportional profit in hand, instead of cashing out, continued to buy and is reported to be sitting on $8 billion in unrealized profits.

The *Forbes* story credits the shale gas revolution with the performance. In a sense that is true, but the explanation is more nuanced than that. First is the fact that the bulk of LyondellBasell profits come from ethylene and derivatives. Second is that ethane prices started to drop relative to oil in 2009. By mid-2010 ethane was nearly half the price of oil on the basis of energy content (see Figure 20). Ethylene is conventionally produced from the oil derivative naphtha. Ethylene produced from cheap ethane is hundreds of dollars cheaper to make per tonne, and a tonne of ethylene sells from $1,000 to $1,500 (the price in 2014) depending on market conditions. This was the primary reason for LyondellBasell's profitability coming strongly out of bankruptcy. But why did other ethylene producers not profit similarly? In chapter 14, I mentioned that 33 of 36 US crackers are on the Gulf Coast, with just two in the Midwest and one small one in Kentucky. Those two in the Midwest just happen to belong to LyondellBasell and are very close to ethane supply. In recognition of this, the company quickly added capacity to these two. Result: enormous profits from capitalizing on the low raw material prices.

In 2013, when Apollo cashed out and Blavatnik increased his stake, ethane pricing dropped through the floor (see Figure 20). This explains why the company profits continued to climb, and the stock stands at 50 percent over the figure when Blavatnik last purchased stock from Apollo. In the *Forbes* interview Blavatnik attributes the success of his strategy to luck and hopes the luck will continue. More likely is the possibility that he saw the trend, which started in 2012. The fundamentals underlying the trend, low natural gas prices and high prices for propane and butane, will continue to cause wet gas to be produced preferentially. Ethane will continue to be a glut until crackers show up. But unlike expansions of LyondellBasell crackers, building new ones takes many more years, significant financing, and ethane pricing crystal balls extending 20 years. Small, distributed crackers, as suggested in chapter 14, could be a shorter-term factor.

The Outlook for Argentina

Argentina is the most promising shale oil and gas play outside the US, even taking into account the EIA estimate in China, where less is publicly known regarding the reservoir character. The combined characteristics of Vaca Muerta make it more attractive than any single play in the US. Shale oil and gas have transformed the US economy, and not just in the energy sector. A net importer of gas is now an exporter. Oil imports have plummeted. A net importer of ammonia fertilizer and of ethylene and derivatives, to name just two classes of chemicals, it will soon be an exporter of these commodities. The effect on the balance of trade is significant. Argentina can aspire to all of this and more. Policy constraints such as the prohibition against export of crude oil are holding back aspects of development in the US. Policy constraints are likely the single biggest class of barriers to a shale revolution in Argentina.

Elephants abound in the room. The nationalization of Repsol-YPF, initially seemingly without compensation, was chilling. The government-majority-owned YPF is doing well, but the chill remains. Until recently, natural gas was allowed to be sold for as low as $2.50 per MM Btu. New production such as shale gas is now permitted a price of $7.50, which is healthy for investors if it holds up. Bolivian gas is being imported for around $10 in mid-2014. LNG is being allowed to be imported at numbers exceeding $13. All of this lends a significant element of uncertainty. Finally, there was the national decision in mid-2014 to default on loans from a US entity. All of the foregoing makes it more difficult to attract financing.

Assuming financing is enabled in some fashion, the highest margin fluids will attract the earliest attention. Even at $7.50 per MM Btu natural gas is not as lucrative as oil at $100 per barrel. At the 2015 price of around $60, things are more in balance. Wet gas is more attractive than dry gas because the wet component sells for close to oil prices and the dry part still has a market. Current estimates have wet gas as a minor part of the Vaca Muerta reserves but with such a lofty numerator even a minor portion is valuable for many years. But if wet gas is produced the ethane will need a home. Conversion to ethylene is attractive compared to production from naphtha, and it would ameliorate the Argentine position as importer of ethylene derivatives. The dry gas in isolation has national energy security value because it could in time eliminate expensive LNG and Bolivian gas. As a net importer of refined products, crude oil, natural gas, and certain petrochemicals, the choice from the smorgasbord above may not matter at first in terms of a national imperative. But at some point certain directions will be more in keeping with national interests as opposed to private commercial ones.

Some directions are crystal clear. The initial YPF/Chevron foray is in the oil leg. The output from this must be refined in the Neuquén province, resisting the temptation to build pipelines to the large refineries in Buenos Aires. This is because this is light sweet oil and needs only a "simple refinery" as defined in chapter 12. One such already exists, the small 25,000 bpd Plaza Huincul refinery in the Neuquén province belonging to YPF. Adding capacity to it would be quicker and less expensive than any other alternative. This is happening in the US today in the Eagle Ford and Bakken.

Plaza Huincul already has gas to methanol conversion capability. This could easily be augmented. In general the federal policy of limiting natural gas availability for methanol production could be relaxed in time. Eventually the country ought to consider legislating allowing methanol as the gasoline oxygenate in place of the current mandate for 5 percent ethanol, especially because half as much methanol achieves the same result. Much as in China, higher concentrations would reduce dependence on gasoline.

Ultimately Argentina can look forward to overall prosperity for every citizen caused by cheap and plentiful domestic energy. In a lecture in Buenos Aires in 2014 I remarked that in the US shale oil and gas had appeared like a tidal wave that lifted all boats of economic prosperity. Much the same is in Argentina's future if policy constraints do not get in the way.

CHAPTER 29

Research Directions

"A new day will dawn for those who stand long"
—From "Stairway to Heaven" by Led Zeppelin (written by Jimmy Page and Robert Plant)

The intrusion of a new reality always brings with it the opportunity for significant research. The sheer newness has associated uncertainties and in many instances also disbelief. This latter was certainly the case with Stanley Prusiner's discovery of prions, at the University of California, San Francisco. The idea of infectious proteins was met with a reaction close to derision from colleagues and bordering on vilification by the scientific press. The prominent journal *Cell* rejected his first paper on the subject. Prusiner and colleagues eventually got it published in the *Proceedings of the National Academy of Sciences*, in principle an even more prestigious journal. But a directed review by a member of the National Academy of Sciences is one mechanism (the member, not an independent editor, picks the reviewers)—when this happens it is not classic peer review.

But publishing in the journal still brings prestige because the members, the crème de la crème of scientific society, are trusted to adhere to scientific rigor in the directed review process. This is sometimes subject to falling prey to impassioned beliefs. Then again, as in this case, groundbreaking science may see the light of day and not be buried by reviewer orthodoxy.

The idea of producing hydrocarbons from a source rock cannot possibly be considered to be in the same league as the discovery of prions. But it shares the attribute that much research activity will be spawned. Unsurprisingly, the major emphasis will be on improving economics of production and minimizing the environmental impact. Beyond this, the sheer abundance and sustained low cost of this resource will drive innovation in the chemical processing and transportation areas. Natural gas fluids will now be an attractive feedstock.

The Prion Story

At the time it was considered impossible that a protein could replicate and that it could transmit disease. Interestingly, the Nobel Prize was awarded to Stanley Prusiner in 1997 before causality was properly established. Criticism for that decision was immense; one science writer likened it to Henry Kissinger's being awarded the Nobel Peace Prize, which is a low blow of major proportions. Basically, people took sides.

Now there is no doubt that bovine spongiform encephalopathy (mad cow disease), Creutzfeldt–Jakob disease, and scrapie (in sheep) are all transmitted by prions. And the case for the dreaded species-to-species transfer has been made.

Even before the solution was seen as groundbreaking, the problem of transmissible spongiform encephalopathy was seen as important enough to merit another Nobel Prize in Physiology or Medicine for its elucidation. In 1976 Carleton Gajdusek was awarded the Nobel Prize. He was studying the disease kuru and hypothesized that the disease was transmitted due to a custom of a particular Papua New Guinea tribe. This essentially involved honoring the death of a family member by eating a portion of that person's brain. The disease existed only among the 65,000 tribal villagers in one set of valleys. Gajdusek also popularized the notion of a slow-acting virus being the cause, and this, too, was considered controversial because no such species had been isolated.

In an interesting parallel to the award to Prusiner, the Nobel was given before definite proof of the hypothesis. But his award spurred an intense search for slow-acting viruses to cure slow-developing illnesses such as cancer and Alzheimer's. The investigators hit a dead end not long after the Nobel award. This is likely the only instance in which a Nobel was awarded first for the wrong explanation for a disease mechanism and then later for the right one.

Improving the Productivity of Wells

This important problem is already being heavily researched by industry. Currently only about 5 to 7 percent of the oil and gas in place is being recovered. This compares to a world average of around 35 percent for conventional reservoirs. Below are just some of my personal favorites regarding directions to be taken.

The steep production decline rate in the first two years or so has been the subject of much conjecture by opponents of shale oil and gas. As noted earlier, this is not a big factor if the overall economics work. But understanding and ameliorating this phenomenon could produce positive returns. One approach already being investigated is that of a proppant that achieves greater penetration than the conventional. This is responsive to the hypothesis that insufficient proppant coverage in the fractures is allowing earth stresses to close the cracks. This progressively chokes off the flow. An extravagant approach would be to throw the conventional book out. No proppants. Figure out some other way to hold the rock faces apart and allow fluid flow. All these measures will be even more important in the case of shale oil, which has less mobility in the fractures than does gas.

I discussed refracturing in chapter 11 and mentioned that there is ongoing research in this area. I have a personal belief that harnessing this may be more fruitful than improving the longer-term flow characteristics of fractures. This is because of the inherently tight nature of the rock. No matter how you improve the conductivity of the fractures, the fact that the rock not immediately proximal is unlikely to drain is something to be dealt with.

An important area is better characterization of the reservoir rock. In many cases up to 35 percent of fracturing stages are believed to be nonproductive. In conventional reservoirs this would essentially be unheard of. Were the rocks better understood, the return on investment would be significantly enhanced. Some of the more interesting work in this area is by a small company named Ingrain using technology (Dvorkin et al., 2003) developed at Stanford University. Here rock core is being evaluated using imaging techniques to provide a quantitative estimate of important reservoir properties such as relative permeability. If the extension of this to evaluating simple drill cuttings is successful, the hit rate on productive intervals ought to improve substantially.

Maybe a combination of technologies is the answer. It will come down to cost-effectiveness. Wet gas, and by extension shale oil, can tolerate a lot of cost if the benefit is there.

Gas to Liquids

Predictably low gas prices combined with the desire to reduce oil imports ought to drive GTL activity. The conventional Fischer-Tropsch process needs improved economics to be able to be competitive with episodic drops in oil price. The North American outlook, and stranded gas in Alaska in particular, should be sufficient impetus for the needed research, most likely in the catalysis aspect. An interesting area for development is that of small-footprint conversion of methane to liquids. This would enable distributed conversion, as mentioned in earlier chapters. But the readily realizable product is a gasoline substitute, diesel being more complicated. The preponderance of gasoline usage in North America for passenger vehicles takes the sting out of this limitation.

The small-footprint conversion of ethane to ethylene is also a target. This is by far simpler in processing terms than the methane conversion mentioned above because ethane already has a carbon-to-carbon bond, and ethylene is simply ethane with two hydrogen atoms removed. Here, too, the principal challenge is innovation in catalysts, although advances will also be needed in the effective separation of constituents, particularly in a portable configuration.

A more mundane but important area is the removal of minor constituents prior to any of this processing. This primarily involves removal of sulfurous gases and carbon dioxide. Alaskan gas sometimes contains up to 12 percent carbon dioxide, although Lower 48 shale gas is advantaged in this regard (with some exceptions). The research again will likely center on small-footprint and minimally energy-intensive methods.

Environmental Issues

These fall broadly in the categories of matters related to water usage and disposal, measurement and control of air emissions, and measures to ensure best practices in all of these.

Water withdrawals: To minimize fresh water usage, saline waters of convenience ought to be employed. The research to tolerate salinity in fracturing fluids has been done to a point. An interesting objective would be to pilot the use of reverse-osmosis reject brine as the base fluid. A quick explanation: reverse osmosis is the current workhorse desalination method. A starting concentration of sea water would end with fresh water and heavy brine. The latter needs to be discarded, and this is problematic in some instances.

The more likely saline water of convenience is from saline aquifers, as discussed earlier. The research needed is in characterizing those proximal to shale reservoirs. At a minimum the constituents evaluated would be bacteria and some of the different salts. A broad effort in each target state is called for. Once characterized and made available to the public, the steps of cleaning and use ought to be straightforward.

Flowback water disposal: In areas with no deep disposal well capability, the problem is significant. All of the Marcellus and Utica for sure have no easy disposal capability, and this has been the cause of serious angst in Pennsylvania. The research will likely focus on the most cost-effective means to reuse the flowback water for fracturing. A lot of work has gone into this aspect, so in the main it will probably be just development activity. Having said that, a better mouse trap could speed the reuse. This probably again will be in small-footprint or portable systems. Large water treatment facilities that aggregate the flowback water from multiple wells will not be the favored route unless there are specific, compelling reasons. Movement of this water in trucks is subject to accidental spills, and the farm roads would welcome less traffic.

Fugitive emissions of methane and, separately, volatile organic compounds have acquired some currency in discussion. While many of these are amenable to simple good practices, there could be scope for better measurement and reporting systems. In April 2012, the US EPA issued a comprehensive set of regulations that primarily address release into the atmosphere. These ought to be actionable by industry without much research. However, the most celebrated of the fugitive methane issues are those of aquifer contamination. Release of detailed plans for testing water wells before drilling (baseline) and at regular intervals during and after drilling is called for. Systems need to be in place to advise homeowners unfamiliar with their rights.

Capturing and utilizing the gas associated with flowback water during the early days of the wells could be a target. To do so economically could be quite challenging. The vented gas onboard LNG ships is utilized effectively because engines exist that can usefully employ the gas. In shale gas operations, switching operating equipment fuel to natural gas would be one target. But that may not consume enough. A high-value target would be the small-footprint conversion of the methane to something useful. Since it will be difficult to make this economic, inducements could be offered to further this sort of endeavor.

Mini LNG: Effective utilization of LNG in transport needs a breakthrough in LNG distribution. LNG is transported at -161°C and is kept cold through controlled evaporative cooling. This limits the range of distribution; there is a use-it-or-lose-it situation. New technology known as mini LNG is in late-stage development. Mini LNG targets production of the liquid in plants rated at a mere 20,000 gallons per day, compared to a typical LNG plant putting out 5 million gallons per day. Such small plants could be distributed all along the transportation network, served by trucks and trains. All they would need is a supply of natural gas. Industrial heavyweights such as Shell, GE, and Linde are reported to be pursuing this market. Mini LNG would also enable the switch away from diesel in other locations such as well site operations.

Social Science Research

Given the national importance of shale gas, research must be conducted on customer perceptions and understanding of the pluses and minuses. Beyond this, communication campaigns ought to be devised to close the gaps in understanding. This will dampen the confusion caused today by stridency from both sides of the debate.

A valid area of research is systematic economic development methodology as it relates to shale gas. In many instances shale gas drilling is in areas new to such industry. Many of the players are small companies that may need guidance in best practices with regard to considering societal benefit in the sustainability equation.

Finally, best practices on casing and cementing need to be detailed, as do methods to track and ensure these practices. Again, this is not the province of research, just something that needs to be done.

CHAPTER 30

Policy Directions

"Come senators, congressmen, please heed the call"

—From "The Times They Are a-Changin'" by Bob Dylan

Shale gas has burst into our energy consciousness with such rapidity as not to allow for a great deal of preparation in the form of informed policymaking. The economic prosperity message has been blunted by the environmental risk cacophony. Both aspects are in need of decided action by all concerned parties in order for the nation to enjoy the economic windfall while still managing the associated risks.

While rulemaking is important, equally critical is voluntary joint industry action. When the industry embarked upon challenging deepwater exploitation, it formed the consortium DeepStar to jointly address the critical hurdles.

The environmental issues facing the shale gas industry do not even begin to approach the difficulty of the deep water tasks. Not to minimize the importance, but with a couple of exceptions, the bulk of the actions require the broad scale execution of best practices. This task can and must be taken on, in part because unconventional resources have, until recently, been the sole province of smaller companies with limited resources.

When the Research Partnership to Secure Energy for America (RPSEA) was formed in 2006 with congressional line-item funding, the two targets were ultra-deepwater and unconventional resources. But even then, when the shale gale was a mere breeze, Congress insisted on a "small operator" provision, meaning that at least some projects were required to have a clear line of sight to benefit the small producer. Today the nation would be well served if RPSEA's focus were to be exclusively tight oil and gas, the latter including shale gas. Strictly from a national imperative, all the evidence now points to tight gas on land being a more important resource than ultra-deepwater gas. The definition of tight gas includes shale gas and gas found in other low-permeability rock such as sandstones and carbonates. Also, with very few exceptions, ultra-deepwater gas is the domain of the big players. They can afford to finance research on their own, or through consortia, such as the recent consortium to devise spill containment technology, in response to the *Deepwater Horizon* disaster.

The use of public funds in implicit support of any major profit-making industry is fraught. Therefore, RPSEA funding of ultra-deepwater efforts is questionable. Risk reduction in shale gas is different. It ought to be considered a national priority because of the economic gain that would otherwise be at risk. Also, the primary beneficiaries, aside from communities in which production happens, will be smaller producers. It is true that this landscape is continually changing, with bigger players buying into the asset base. But the charge that support of shale gas is support of Big Oil has little merit in the main. The list of producers in Pennsylvania today is a Who's Who of companies that the public never heard of. Ironically, the entrée of bigger players is likely to improve operational diligence. Some may not like the size and profitability of ExxonMobil, but for operational discipline they have no equal.

The ideas expressed below are intended to stimulate discussion and not be overly prescriptive. The most important areas are those that ensure environmentally secure operations. Within each category, though, my ideas are addressed in no particular order.

Environmental Issues

Preventing well water contamination: The most important measure is to ensure the collection of baseline data on a specified set of chemicals including methane in all wells in a predefined proximity to the drilling operations, likely 2,500 feet. The appropriate authority (federal, state, or county) should specify the nature and frequency of the testing and qualify the testing entities. The cost ought to be borne directly or indirectly by the producer. The purpose is to definitively assess whether fugitive methane or any other species from the production operations are entering well water. In the event such a leak is detected, regulations ought to be in place for remedial action to fix the leak.

Well operators should be required to place cement sealing over every zone of potential gas production, no matter how small in quantity, above the reservoir. Failure to do so incurs the risk of a source for leakage. This cement job and the normal casing and cementing operations ought to be verifiably competent, as determined by appropriately specified testing. Industry best practice in this area should be documented and made available to all producers. Regulatory oversight for compliance may be needed. This can be facilitated by the use of sophisticated sensors and means to transmit via satellite the data to operational supervisors as well as regulators in central urban locations, thus minimizing oversight cost. This last is very simply accomplished and is commonplace in the industry today. But the driver for remote communications has been operational cost reduction, especially for deepwater operations. Adapting to measurements required by regulators would be simple. The suggestion is that an industry-sponsored organization coordinate this effort and ensure that both small operators and regulators have remote communication (satellite) capabilities. Every state is writing new regulations in this area, which are increasingly prescriptive. We need to throw technology at the problem of cost of monitoring and enforcement. Failing to do so will overload the state departments of environmental protection and siphon away too great a fraction of the severance tax.

Tax matters: Nobody likes new taxes. But severance taxes are fair because they are philosophically no different from corporate income taxes. In addition to this, all states ought to consider impact fees. These are fees to directly address local infrastructure damage such as to roads and bridges from the unusually heavy traffic associated with fracturing operations. The proceeds from these fees ought to remain with the local community suffering the privation.

Handling flowback water: Regulations must be in place for safe handling of flowback water. The most preferred options are to treat and reuse, and disposal in UIC Class II wells. Consideration could be given to making mandatory only these two options, with variances granted only on a case-by-case basis. Disposal in UIC Class II wells is less expensive than treating and reusing flowback water when available. However, further diligence on such wells relative to procedures to minimize the risk of interaction with active faults is needed. Failure to do so can lead to minor earthquakes. Reusing flowback water will be facilitated by a practice of using salty water as the base fluid for fracturing. Even so, service providers will need to gear up to provide the treatment service, so benign alternatives may have to be found for each area in the interim.

Minimizing fresh water usage: Industry today can feasibly utilize salty water for fracturing. Consideration should be given to regulations requiring this as the default, with explainable variances only in cases where salty water of convenience is absent or temporarily unavailable.

Disclosure of fracturing chemicals: This issue is more fully discussed in chapter 6, "The Chemicals Disclosure Conundrum." The proprietary product exception must not be permitted to be used as a shield or an artifice. One measure to assure that is to require a state body to verify the claim for exemption. The nonprofit website FracFocus is now almost a standard of use as a simple vehicle for disclosure and public access.

Minimizing fugitive gas emissions: Gas produced prior to the hooking up of a pipeline to the production site must be handled in the most environmentally secure fashion. In chapter 18, I recommend innovation in small-footprint, preferably mobile, means to capture the gas and convert it to a transportable liquid. For economics purposes this gas could be considered to be zero cost. If the new EPA green completions regulations slated for 2015 go through unscathed, methane release and flaring will not be permitted for new operations. The associated gas from oil production is not currently covered by the legislation, which also has other exemptions. But technologies are needed to address legacy and currently exempted operations.

Displacing diesel use: The biggest energy hogs on shale oil or gas operations are the high-pressure pumps associated with fracturing. These operate on diesel. In close proximity to farms and residences, this constitutes a hazard relative to particulate emissions. $PM_{2.5}$ (particulate matter smaller than 2.5 microns in size) is a scourge and is responsible for 1.5 million preventable deaths worldwide. Diesel can be substituted with natural gas alternatives such as CNG or LNG. Pilot operations are under way, with Apache Corporation leading the way with the help of equipment manufacturer Caterpillar and service provider Halliburton. Other entities have since climbed on board. A key to success is the economic supply of CNG or LNG at the rig site. This is discussed in chapter 29, "Research Directions."

Pad drilling: This is a mode of production wherein multiple wells are drilled from a single location. This facilitates all of the recommendations made above and has a positive environmental footprint with respect to roads and traffic. It is a net benefit for the producer as well because a few experienced operators can now supervise several wells. Best practices must be shared to nudge pad activities to commence as early in the development as possible.

Economic Issues

Ethane monetization: State and local governments should consider inducements for local processing of the ethane, be they regional crackers or pad-level reactive options (see chapter 14). The compelling economics may be sufficient, but the governments should also take full responsibility for local workforce training and the like. Consideration ought also to be given to encouraging research and development centers in the Marcellus/Utica region targeting research related to wet gas. Proximal production would allow easy access to field testing sites.

Export issues: Abetted by a warm winter, abundant shale gas caused the price in early 2012 to be the lowest in a decade. Even if this is ignored as aberrant, there is little doubt that a lot of cheap gas will become the norm. In fact, a colder winter in 2013–2014 caused the prices to revert to the new normal of around $4, with minor excursions to $6. This last was entirely due to the inadequacy of midstream infrastructure and was short-lived. The temptation to export the gas is high, and this is currently being debated. The box below gives some views on this matter.

Unconstrained LNG Export Is Not in the National Interest

We must minimize the export of natural gas in any form in favor of producing and exporting a higher-value product. The single most valuable such high-volume product is ammonia-based fertilizer. (Carbon black would be higher value but has a smaller market.) Until recently the US imported half the fertilizer consumed. This is because variable and high prices in the early part of the century caused many manufacturers to relocate abroad to areas of cheap gas such as the Middle East. Now with the prospect of cheap and stable shale gas, many of these are returning. No doubt the chemical industry is skittish about LNG export concepts because exports could vitiate the business assumptions of low cost, were the prices to rise due to massive export of gas. At the time of writing, five permits have been issued, totaling about 10.5 bcf per day of gas usage. An additional 5 bcf per day could be permitted without a material impact on gas prices. Maintaining the price differential between North America and Europe (and Asia) is key to the expected manufacturing renaissance in the chemicals and fuels sector.

Aside from the pricing issue, another reason to export product rather than gas is simple economics. Take the example of anhydrous ammonia, the basic building block for nitrogen fertilizer manufacture. About 33.3 mcf gas converts to 1 tonne of anhydrous ammonia. The gas value, using $4 per mcf, is $134. The value of the anhydrous ammonia is in the vicinity of $800. Also, domestic labor was used to get it to that state. The landed price in Europe of gas as LNG will be about $8.50 with $4 gas. That near-doubling of value added does not contribute much to the domestic economy. Even the ship was probably made in Korea.

The US will be one of the lowest-cost producers of ethane-based ethylene and derivative polymer in the world. This raises the high likelihood of the US being a net exporter of these chemicals. If gas prices stay low, wet gas will be produced almost exclusively. This could cause an ethane glut and lead to exports of ethylene derivatives. This ought to be permitted and encouraged.

States proximal to wet gas production but without any of their own, such as North Carolina, ought to consider encouraging chemical production. Wilmington, North Carolina, already has a chemical industry and being a port could be an export site. Prosperity from cheap and abundant gas does not have to be restricted to the producing states.

Exporting shale oil: As explained in chapter 12, domestically produced light sweet oil from shale reservoirs cannot find a ready market in the US due to domestic refineries' being geared to heavy oil. Consequently it is selling at a discount to the world market price of the commodity. In January 2014, the discount was $18 less than WTI and $27 less than Brent. One could expect the export price to be between the two. From a national standpoint, if limited export is permitted and the shortfall made up with extra Canadian crude, there ought to be a net positive impact on GDP. This is because Canadian heavy oil sells between $15 and $30 less than WTI. Selling high and buying low has always been a good strategy!

Oil substitution: Three principal avenues present themselves: electric vehicles, natural gas replacement of gasoline and diesel, and conversion of natural gas to liquids. The potential policy drivers for this could include the following:

- A decision by Alaska to kill the gas pipeline to the Lower 48 and to encourage the conversion of vast quantities of cheap stranded gas to liquids. The owners of the gas, principally major oil companies with a thorough understanding of GTL technologies, ought to be motivated to do this with no inducement. Absent this, the TAPS, the pipeline bringing oil down from the North Slope, is at risk of closure, as discussed in chapter 15.

- Shale gas–producing states should consider requiring operators to displace diesel with CNG, LNG, DME, or methanol for the pressure pumps used in fracturing and cementing, and natural gas or methanol for the vehicles. There will be issues of access to refueling, capital cost of retrofits, and the like, but states could address the refueling issues, at least. This could be a bit tricky because the fuel of choice for trucks would be LNG, not CNG.

- All metropolitan areas ought to consider emulating Delhi and a host of other Indian cities where all public transport switched to CNG. The World Bank evaluated the health benefits as very significant, as noted in chapter 17. But from the standpoint of reducing imported oil, replacing gasoline has a bigger bang and methanol may be the route, as discussed in chapter 19.
- Dimethyl ether (DME) can substitute for diesel up to at least 20 percent with no engine modification, emits zero particulates, and has a very high cetane rating. Cheap natural gas equates to cheap DME. States should consider DME additive to diesel as an alternative to a wholesale switch to natural gas or as an early step toward that goal. New plants for DME production will be required. Since DME is a single processing step beyond methanol production, methanol and DME strategies can be pursued simultaneously.

National Security

The military is a gigantic user of transport fuel. Relative to fuel switching, on the one hand mission criticality would demand the most proven and reliable fuel. On the other hand, the delivery of such a fuel to the front lines has huge elements of cost and potential loss of life. The military ought to conduct an analysis of options, some considerations of which are listed below:

- Base operations are amenable to a complete switch to CNG-fueled vehicles. In fact it may not be far-fetched for military vehicle design to take advantage of methane's 125+ octane rating with a high-compression engine. If the military standardized on such an engine, a civilian version down the road could be feasible. Think Hummer.
- Bases, particularly in foreign countries, could operate small-scale electricity generators using natural gas with stored backup of CNG or even methanol. This could be augmented with solar power in appropriate locations. Nonreliance on an outside grid would have security advantages.
- Aside from the physical risks in delivering fuel and water to the front lines, the cost of security for the convoy likely raises the cost of these commodities to many times the normal. Consequently, the military could consider sponsoring research in the small-footprint distributed production of transport fuel using the raw material of convenience. The baseline cost for such fluid would be the security-inflated cost of convoy

delivery. Also, a switch to electricity for vehicles where feasible should be strongly considered. Electricity production in forward locations is more feasible than liquid fuel production.

Finally, abundant shale gas in North America means that more LNG is available for delivery to Europe, thus significantly reducing Russia's ability to use gas supplies as a weapon of political will. One could expect increased US influence worldwide as a result. While no specific policy actions spring to mind, lawmakers ought to be aware of the consequences in this regard of significantly holding back shale gas production in the US.

Conclusions

"There will be an answer, let it be"

—From "Let It Be" by The Beatles (written by John Lennon and Paul McCartney)

Shale gas has the potential to materially improve the economic lot of every citizen of the US. There is also the real possibility that, together with distributed cleaner energy such as wind and solar, cheap energy in the form of shale gas could improve the human condition worldwide. Such lofty rewards beg for a concerted effort by all concerned to assess and ameliorate the associated environmental risks. In the opinion of this author, that is an acceptable risk provided we employ a combination of innovative technology, regulatory oversight, and the industry will to do the right thing, nudged along by informed local activism in the areas that industry operates.

The public has a right to know the timeframes involved in the rewards and the risks. While the intent of the book was to allow the readers to form their own opinions, I will attempt a response to the question of timing. The reduction in the cost to heat homes was already felt in the winter of 2011; one estimate is a $1,000 average reduction per household each year. In 2012 this number was estimated to be $1,200. Low-cost fertilizer and ethylene and derivatives are already making their presence felt from the standpoint of reduced imports and will do so increasingly in the short term: two to five years. The quicker results will be from resuscitating previously mothballed plants. To the extent that there is a world price for these items, export could dampen the direct benefit to farmers and other consumers. Policy measures could ensure the domestic benefit. In any case the balance of trade would improve if for no other reason than through the reduction, or in the limit elimination, of imports. In early 2014 the American Chemical Council reported that new shale gas–related capital planned totaled $100.2 billion. It appears that nearly half of this funding is from foreign sources, underlining the fact that the chemical industries in other countries are non-competitive due to cheap North American gas.

Methanol displacement of gasoline and diesel is an exciting consequence of shale gas production. The needed legislation to require most cars to accept any combination of conventional fuels and alcohol still awaits passage in Congress. Meanwhile China is leading the way, although the methanol is entirely being sourced from coal, not natural gas. That could change with the anticipated supply of gas from Russia, reported to be in the vicinity of 1.4 trillion cubic feet per year. The compelling economics could drive change in two to five years in the US. The slowest link in that chain will be addition of methanol capacity and establishing fueling infrastructure. The high compression engine which effectively would more than halve the cost of fuel per mile will take longer: up to 10 years. This timeframe could move up if the public is vociferous in demanding it.

The positive effects on US national security for dimming the aspirations of Russia and Iran would be in the short time frame, two to five years. It would be longer for the effect on reduction of imported oil, closer to the 10-year mark for a material effect. However, the most recent surge in shale oil production is affecting this assessment directly, and the trend is expected to continue. There is a realistic possibility that by 2025 all oil will be sourced from North America.

On the environmental risk side, the public will want to know whether measures to manage the risk are working and how soon this will be known. Regulations requiring that flowback water only be either reused or disposed of in UIC Class II wells ought to be in place by 2015, if states choose to impose them. North Carolina, the newest state to formulate regulations, expects to have these in place by early 2015. But regulations in and of themselves do not assure compliance.

The measures to assure prevention of earthquakes from disposal well activity ought to be a short term activity. Proper measures, such as the requirement to map the subsurface for proximity to active faults, ought to eliminate this phenomenon. The appropriate place for this is in modification of the specifications in EPA's UIC Class II well procedures. Technology for reuse of flowback water already exists but more will be needed for broadening use and reducing cost. This is the more desirable of the two options. Full-scale implementation ought to be possible in two to five years. Since programmed releases of flowback water to the surface will not be an option, any accidental spills during the two allowable handling methods would be observable events.

With the industry trending toward the use of greener constituents in fracturing fluid, the only substantive issue with accidental spills should be the saltiness. Environmental degradation just from high salt would not be as long lasting as it might have been from organic compounds such as diesel, which would be forbidden from use in that application in most states. There is no technical need to use diesel in fracturing fluid, and disallowing it does not constitute restraint of trade.

The use of salty water in place of fresh for fracturing is technically feasible today. Material implementation ought also to be in the short term provided there is the will to do it. State legislation requiring this would be desirable and community activism would be helpful in persuading legislators.

Preventing well water contamination ought to be straightforward if my suggestions in this book, and available in greater detail in the North Carolina regulations promulgated in early 2015, are followed. A key aspect is that industry must make best practices on well construction and monitoring available to the smaller producers. Baseline testing of proximal water wells at industry expense prior to drilling activity, followed by routine testing after, will provide definitive proof of efficacy of well construction. There would no room for uncertainty. In the event of an upset condition, response time for remediation ought to be weeks not months. From an operational standpoint, sound well construction preventing fluid leakage is just a matter of following good practice. Consequently, upset conditions ought to carry severe penalties.

Methane releases during early stages of gas production are believed to occur in some instances. In each case the gas is first collected, so the operator knows how much there is and precisely where it is stored. This potent greenhouse gas must not be directly discharged. If no useful purpose is found, it ought to be burned on location (flared). If this gas is either used or flared, the public ought not to have concerns on this score. If operators are required to report on these occurrences, an even greater measure of comfort could be realized without undue burden on the operator. Automatic monitoring with remote reporting ought to be straightforward. Provided the so-called green completions mandate by the EPA is fully enforced in 2015, all the foregoing is moot. No discharge or flaring will be permitted for new production, although there are exemptions for exploratory wells, and gas associated with oil production is also not covered at this time. The study by the University of Texas (Allen et al., 2013), funded largely by the Environmental Defense Fund, is endeavoring to detail precisely the nature and quantity of fugitive emissions. When completed it ought to provide direction on how best to ameliorate the problem.

There is little doubt that shale gas is transforming the US energy-based economy. An importer of natural gas, with 10 percent imports in 2010, we are essentially already in an era of domestic self-sufficiency in 2014. Now the argument has shifted to how much export of liquefied natural gas is in the national interest. The sheer abundance of gas is keeping the prices low, and models from two different sources indicate that the prices should remain low to moderate for decades. An important aspect of note is that the prices can be expected to remain stable, without the peaks and valleys experienced in the past. This stability is almost as much an attraction to industrial users as the low pricing.

Even at the upper end of the modeled pricing range, $8 per MM Btu, gas-based electricity will be cheaper than that from new coal plants, even without a price on carbon. But if prices stay in the lower end of the range for long, renewable energy sources will be slow to be adopted unless there is policy intervention. Another consequence of sustained prices at the low end will be emphasis on wet gas production, because the majority of the profit will be in the wet component. This will cause a glut in ethane, which, if properly anticipated through cracker capacity addition, could make the US the low-cost producer of ethylene, allowing for lucrative exports.

The explosion in shale oil production will eventually dampen the ability of OPEC to manipulate prices. In the shorter term, the light sweet oil is not permitted to be exported, so the effect is felt through the diverting of that type of oil away from the US to other markets. The oil export issue will be strongly debated in 2015, especially with the Republican majority in Congress. The most recent plummeting of oil prices in late 2014 underlines the need to allow exports. Absent exports, domestic production will in effect have a price at least $6 lower than world prices. This will reduce the breakeven price for US production, thus adversely affecting jobs.

Low cost energy is a tide that lifts all boats of economic growth. Shale gas is a powerful such tide. It has burst upon us so unexpectedly that we have become rattled by the flotsam it carried with it. This author concludes that the flotsam is manageable, allowing us to enjoy the benefits of the tide.

Bibliography

These represent citations made in the text plus other suggested reading. Inevitably many of the sources are news stories and blogs. In using the information from each of these, I have considered the source for possible slant or bias. But readers ought to draw their own conclusions in this regard. Many of the assertions made in the essays are drawn from my experience. In these cases no citations are made.

Akbar, S., Lvovsky, K., Kojima, M., & World Bank South Asia Region. (2005, June 1). For a breath of fresh air: Ten years of progress and challenges in urban air quality management in India, 1993–2002 (English). Working paper 35047. New Delhi: World Bank Environment and Social Development Unit, South Asia Region. Retrieved March 19, 2012, from http://documents.worldbank.org/curated/en/2005/06/6561544/breath-fresh-air-ten-years-progress-challenges-urban-air-quality-management-india-1993-2002

Allen, D. T., Torres, V. M., Thomas, J., Sullivan, D. W., Harrison, M., Hendler, A., ... Seinfeld, J. H. (2013). Measurements of methane emissions at natural gas production sites in the United States. *Proceedings of the National Academy of Sciences (PNAS), 110* (44): 17768–17773. Available from http://www.pnas.org/content/110/44/17768.full.pdf

Alyeska Pipeline Service Co. (2010). *Trans Alaska Pipeline System; The facts.* Retrieved July 29, 2015, from http://www.alyeska-pipe.com/assets/uploads/pagestructure/NewsCenter_MediaResources_FactSheets_Entries/635078372894251917_2013AlyeskaTAPSFactBook.pdf

American Chemistry Council. (2014, February 20). *US chemical investment linked to shale gas reaches $100 billion* [news release]. Washington, DC: American Chemistry Council. Retrieved May 7, 2014, from http://www.americanchemistry.com/Media/PressReleasesTranscripts/ACC-news-releases/US-Chemical-Investment-Linked-to-Shale-Gas-Reaches-100-Billion.html

Ariely, D. (2008). *Predictably irrational: The hidden forces that shape our decisions*. New York: Harper Collins.

Barcella, M. L., Gross, S., & Rajan, S. (2011). *Mismeasuring methane: Estimating greenhouse gas emissions from upstream natural gas development*. Cambridge, MA: IHS CERA. Retrieved from https://www.heartland.org/sites/default/files/Mismeasuring%20Methane.pdf

Benchaar, C., Pomar, C., & Chiquette, J. (2001). Evaluation of dietary strategies to reduce methane production in ruminants: A modelling approach. *Canadian Journal of Animal Science, 81*, 563–574. http://dx.doi.org/10.4141/A00-119

Bromberg, L., & Cohn, D. R. (2008). *Effective octane and efficiency advantages of direct injection alcohol engines*. MIT Laboratory for Energy and the Environment, Report LFEE 2008-01 RP.

Brune, M. (2012, February 2). The Sierra Club and natural gas [Web log post]. Retrieved May 14, 2012, from http://sierraclub.typepad.com/michaelbrune/2012/02/the-sierra-club-and-natural-gas.html

Brusstar, M., Stuhldreher, M., Swain, D., & Pidgeon, W. (2002). *High efficiency and low emissions from a port-injected engine with neat alcohol fuels*. SAE Technical Paper 2002-01-2743. Washington, DC: US Environmental Protection Agency.

Cathles, L. M., Brown, L., Hunter, A., & Taam, M. (2012, February 29). Press release: Response to Howarth et al.'s reply. *Cornell Geological Sciences*. Retrieved March 14, 2012, from http://www.geo.cornell.edu/eas/PeoplePlaces/Faculty/cathles/Natural%20Gas/Response%20to%20Howarth's%20Reply%20Distributed%20Feb%2030,%202012.pdf

Chelme-Ayala, P., El-Din, M. G., Smith, R., Code, K. R., & Leonard, J. (2011). Advanced treatment of liquid swine manure using physico-chemical treatment. *Journal of Hazardous Materials, 186*(2–3), 1632–1638.

Chesapeake's new Utica Shale wells producing "very strong." (2011, September 29). *Marcellus Drilling News*. Retrieved March 19, 2012, from http://marcellusdrilling.com/2011/09/chesapeakes-new-utica-shale-wells-producing-very-strong/

Cohn, D. R. (2012, March 27). *Use of methanol in flex fuel heavy duty trucks*. Presentation at Methanol Policy Forum, Washington, DC.

Considine, T. J., Watson, R., & Blumsack, S. (2011). The Pennsylvania Marcellus natural gas industry: Status, economic impacts and future potential. State College, PA: Pennsylvania State University College of Earth and Mineral Sciences, Department of Energy and Mineral Engineering.

Dayton, D. (2013, September 3–6). *Catalytic biomass pyrolysis technology development for advanced biofuels*. Presentation at tcbiomass2013, the Third International Conference on Thermochemical Conversion of Biomass, Chicago, IL. Retrieved June 13, 2014, from http://www.gastechnology.org/tcbiomass2013/tcb2013/11-Dayton-tcbiomass2013-presentation-Thur.pdf

Despite methane emissions upstream, natural gas is cleaner than coal on a life-cycle basis. (2011, August 26). Retrieved March 15, 2012, from Worldwatch Institute website: http://www.worldwatch.org/despite-methane-emissions-upstream-natural-gas-cleaner-coal-life-cycle-basis

Dvorkin, J., Kameda, A., Nur, A., Mese, A., & Tutuncu, A. N. (2003, May). Real time monitoring of permeability, elastic moduli and strength in sands and shales using Digital Rock Physics. Presented at the Society of Petroleum Engineers European Formation Damage Conference, The Hague, Netherlands. http://dx.doi.org/10.2118/82246-MS

The economics of LNG. (n.d.). Retrieved from Shareholdersunite.com website: http://shareholdersunite.com/the-ioc-files-useful-background-material/the-economics-of-lng/

Economides, M. (2005). The economics of gas to liquids compared to liquefied natural gas. *World Energy, 8*(1), 136–140. Retrieved March 14, 2012, from http://www.worldenergysource.com/articles/pdf/economides_WE_v8n1.pdf

Europeans shiver as Russia cuts gas shipments. (2009, January 7). *NBCNews.com*. Retrieved June 16, 2014, from http://www.nbcnews.com/id/28515983/ns/world_news-europe/t/europeans-shiver-russia-cuts-gas-shipments/

Fleisch, T. (2014). Associated gas monetization via miniGTL: Conversion of flared gas into liquid fuels & chemicals, April 2014 update. Prepared for the World Bank/Global Gas Flare Reduction Partnership. Retrieved April 2014 from http://www.infratechnology.ru/media/documents/2014/04/miniGTL_Update_report_1.pdf

The future of biofuels: The post-alcohol world. (2010, October 28). *The Economist*. Retrieved March 15, 2012, from http://www.economist.com/node/17358802

Go figure: A carbon tax crafted right here at home. (2007, March 9). *Calgary Herald*. Retrieved from http://www.canada.com/calgaryherald/columnists/story.html?id=8c3c9760-7cbe-4fab-b00c-1c77243903b6

Gold, R., & Campoy, A. (2011, December 6). Focus on fracking: Oil's growing thirst for water. *The Wall Street Journal*. Retrieved March 12, 2012, from http://online.wsj.com/article/SB1000142405297020452820457700993022284 7246.html

Halliburton. (2015). Fluids disclosure. Retrieved July 9, 2015, from http://www.halliburton.com/public/projects/pubsdata/Hydraulic_Fracturing/fluids_disclosure.html

Harrison, M. (2012, January 17). Revised attachment 3: Gas well completion emissions data, URS Corporation report. [Attachment to memo from A. Farrell & B. Thompson to G. McCarthy, US Environmental Protection Agency]. America's Natural Gas Alliance. Retrieved June 4, 2015, from http://anga.us/media/testimony/5A845E4B-5056-9F69-D41C3DB6726FCA46/files/anga%20axpc%20nsps%20neshap%20comments.pdf

Heintz, J., & Pollin, R. (2011). *The economic benefits of a green chemical industry in the United States: Renewing manufacturing jobs while protecting health and the environment*. Political Economy Research Institute, University of Massachusetts, Amherst. Retrieved March 14, 2012, from http://www.hhh.umn.edu/centers/stpp/research/pdf/UMassGreen-Chemistry-Report_FINAL.pdf

Howarth, R. W., Santoro, R., & Ingraffea, A. (2011). Methane and the greenhouse-gas footprint of natural gas from shale formation. *Climatic Change, 106*(4), 679–690. http://dx.doi.org/10.1007/s10584-011-0061-5

International Energy Agency (IEA). (2012). *Energy security*. Retrieved May 21, 2012, from http://www.iea.org/topics/energysecurity/subtopics/whatisenergysecurity/

Jackson, R., & Vengosh, A. (2011, May 10). Strong evidence that shale drilling is risky. *Philly.com*. Retrieved March 15, 2012, from http://articles.philly.com/2011-05-10/news/29528421_1_water-wells-safe-drinking-natural-gas

Johnson, J. (2012, March 19). A long view of the energy market. *Chemical & Engineering News, 90*(12), 34–35.

Kaufman, L. (2011, December 17). Environmentalists get down to earth. *The New York Times*. Retrieved March 19, 2012, from http://www.nytimes.com/2011/12/18/sunday-review/environmentalists-get-down-to-earth.html

Kindy, K., & Keating, D. (2008, November 23). Problems plague US flex-fuel fleet. *Washington Post*. Retrieved March 15, 2012, from http://www.washingtonpost.com/wp-dyn/content/article/2008/11/22/AR2008112200886.html

King, G. (2011, May). A closed loop system using a brine reservoir to replace fresh water as the frac fluid source. In *Proceedings of the Technical Workshops for the Hydraulic Fracturing Study: Water Resources Management* (EPA 600/R-11/048; pp. 76–80). Washington, DC: US Environmental Protection Agency.

King, G. (2012). *Hydraulic fracturing 101: What every representative, environmentalist, regulator, reporter, investor, university researcher, neighbor and engineer should know about estimating frac risk and improving frac performance in unconventional gas and oil wells*. Society of Petroleum Engineers conference paper 152596-MS. http://dx.doi.org/10.2118/152596-MS

Kresse, T., Warner, N., Hays, P., Down, A., Vengosh, A., Jackson, R. (2012). Shallow groundwater quality and geochemistry in the Fayetteville Shale gas-production area, north-central Arkansas, 2011: US Geological Survey Scientific Investigations Report 2012–5273. Little Rock, AR:US Geological Survey.

Levi, M. (2011, April 15). Some thoughts on the Howarth Shale gas paper [Web log post]. Retrieved March 14, 2012, from Council on Foreign Relations website: http://blogs.cfr.org/levi/2011/04/15/some-thoughts-on-the-howarth-shale-gas-paper/

Luft, G., & Korin, A. (2009). *Turning oil into salt: Energy independence through fuel choice*. Charleston, SC: Booksurge Publishing.

Marron, D. (2009, August 21). The disconnect between oil and natural gas prices [Web log post]. Retrieved May 16, 2012, from http://dmarron.com/2009/08/21/the-disconnect-between-oil-and-natural-gas-prices

Martin, C., Ferlay, A., Chilliard Y., & Doreau, M. (2007, September). Rumen methanogenesis of dairy cows in response to increasing levels of dietary extruded linseeds. In *2nd International Symposium on Energy and Protein Metabolism and Nutrition*, 9–13 September 2007, Vichy, France, pp. 609–610.

Medlock, K., Jaffe, A., & Hartley, P. (2011). *Shale gas and US national security. Energy forum*. Houston, TX: James A. Baker II Institute for Public Policy, Rice University. Retrieved June 4, 2015, from http://bakerinstitute.org/research/shale-gas-and-us-national-security/

Michaud, K., Buccino, J., & Chenelle, S. (2014, March 14). The impact of domestic shale oil production on US military strategy and its implications for US-China maritime partnership. *Small Wars Journal*. Retrieved May 7, 2014, from http://smallwarsjournal.com/jrnl/art/the-impact-of-domestic-shale-oil-production-on-us-military-strategy-and-its-implications-fo

Molofsky, L., Connor, J., Farhat, S., Wylie, A., Jr., & Wagner, T. (2011, December 5). Methane in Pennsylvania water wells unrelated to Marcellus shale fracturing. *Oil & Gas Journal*, 54–67. Retrieved March 14, 2012, from http://www.cabotog.com/pdfs/MethaneUnrelatedtoFracturing.pdf

Mueller, K., & Yeston, J. (Eds.). (2011, December 2). *Don't panic* [Editors' Choice]. *Science, 334*, 1183. http://dx.doi.org/10.1126/science.334.6060.1183-a

Muller, R. A., & Muller, E. A. (2013, December). Why every serious environmentalist should favour fracking. London: Centre for Policy Studies.

National Energy Technology Laboratory (NETL). (2009, April). *Alaska North Slope oil and gas: A promising future or an area in decline?* Addendum report DOE/NETL-2009/1385. US Department of Energy. Retrieved June 4, 2015, from http://www.netl.doe.gov/file%20library/Research/oil-gas/ANS_Potential.pdf

Nikhanj, M., & Jamal, S. (2010, March 23). No more guessing: Gas rig efficiency and spud to sales. *Natural Gas Supply*. Calgary: Ross Smith Energy Group.

Norwegian Petroleum Directorate. (2010, July 5). Environmental considerations in the Norwegian petroleum sector. In *Facts 2010*. Retrieved June 4, 2015, from http://www.npd.no/en/Publications/Facts/Facts-2010/

Oil and natural gas sector: New Source Performance Standards and National Emission Standards for Hazardous Air Pollutants Reviews. US Environmental Protection Agency (EPA). 40 C.F.R. Part 63 (2012). http://www.epa.gov/airquality/oilandgas/pdfs/20120417finalrule.pdf

Olah, G., Goeppert, A., & Surya Prakash, G. (2009). *Beyond oil and gas: The methanol economy*, 2nd ed. Weinheim, Germany: Wiley-VCH.

Osborn, S. G., Vengosh, A., Warner, N. R., & Jackson, R. B. (2011). Methane contamination of drinking water accompanying gas-well drilling and hydraulic fracturing. *Proceedings of the National Academy of Sciences, 108*(20): 8172–8176.

PFC Energy. (2009, July). *Unpacking uncertainty: Investment issues in the petroleum sector*. Retrieved from International Energy Forum website: http://www.ief.org/_resources/files/events/12th-ief-ministerial-cancun-mexico/unpacking-uncertainty-full.pdf

Phillips, M. (2013, January 16). Falling US oil imports will reshape the world crude market. BloomburgBusinessweek.com. Retrieved from http://www.businessweek.com/articles/2013-01-16/falling-u-dot-s-dot-oil-imports-will-reshape-the-world-crude-market

Plastics prices: Pretty pricey polymer. (2011, February 14). *The Economist*. Retrieved March 15, 2012, from http://www.economist.com/blogs/newsbook/2011/02/plastics_prices

Rao, V. (2011, June). Gas positioned to displace coal plants. *The American Oil & Gas Reporter*. Retrieved June 4, 2015, from https://rtecrtp.files.wordpress.com/2010/06/aogr-article.pdf

Rao, V. (2014, April). Shale gas drives vertical integration. Guest editorial. *JPT*, 18–19.

Rodvelt, G., Yeager, V., & Hyatt, M. (2011, August). *Case history: Challenges using ultraviolet light to control bacteria in Marcellus completions* (SPE 149445). Prepared for presentation at the Society of Petroleum Engineers Eastern Regional Meeting, Columbus, OH. Retrieved January 13, 2015, from http://www.baroididp.com/premium/tech_papers/source_files/pe/H08643_SPE149445.pdf

Royal Dutch Shell plc. (2014, September 3). *Shell makes Utica gas find in Pennsylvania* (press release). Retrieved January 13, 2015, from Rigzone website, http://www.rigzone.com/news/oil_gas/a/134839/Shell_Makes_Utica_Gas_Find_in_Pennsylvania

Ryder, R. T. (2008). *Assessment of Appalachian Basin oil and gas resources: Utica-Lower Paleozoic total petroleum system*. Open File Report 2008-1287. Reston, VA: US Geological Survey.

Seddon, D. (2010). *Petrochemical economics: Technology selection in a carbon constrained world*. London: Imperial College Press.

Sharma, M. (2010). *Improved reservoir access through refracture treatments in tight gas sands and gas shales* [Presentation]. Retrieved March 14, 2012, from Research Partnership to Secure Energy for America website, http://www.rpsea.org/media/files/project/90d3f4fe/EVNT-PR-07122-41_2010_Improved_Reservoir_Access_Refracture_Treatments_Tight_Gas_Sands_Shales_Sharma-04-06-10.pdf

Simmons, M. R. (2005). *Twilight in the desert: The coming Saudi oil shock and the world economy*. Hoboken, NJ: John Wiley & Sons.

Somers, J., & Schultz, H. (2010). Coal mine ventilation air emissions: Project development planning and mitigation technologies. In S. Hardcastle & D. McKinnon (Eds.), *13th United States/North American Mine Ventilation Symposium*. Sudbury, Ontario: MIRARCO. Retrieved March 14, 2012, from the US Environmental Protection Agency website, http://www.epa.gov/cmop/docs/vam-planning-mitigation.pdf

Switching from coal to natural gas would do little for global climate, study indicates. (2011, September 8). Retrieved March 15, 2012, from http://www.physorg.com/news/2011-09-coal-natural-gas-global-climate.html

Thaler, R. H., & Sunstein, C. R. (2003). Libertarian paternalism. *American Economic Review, 93*(2), 175–179.

Turner, T. (2011, December 6). Fracking deal is reached by YPF, Petronas. *The Wall Street Journal*. Retrieved November 7, 2014, from http://online.wsj.com/articles/fracking-deal-is-reached-by-ypf-petronas-1409209249

Turner, T. (2014, November 14). Why Argentines are unmoved by sky-high inflation [Web log post]. *The Wall Street Journal*. Retrieved January 23, 2015, from http://blogs.wsj.com/frontiers/2014/11/03/why-argentines-are-unmoved-by-sky-high-inflation/

Urbina, I. (2011, June 25). Insiders sound an alarm amid a natural gas rush. *The New York Times*. Retrieved March 14, 2012, from http://www.nytimes.com/2011/06/26/us/26gas.html?_r=1&pagewanted=1&sq=marcellus&st=cse&scp=12

US Energy Information Administration (EIA). (2010, January 27). *Schematic geology of natural gas resources.* Retrieved May 22, 2012, from http://www.eia.gov/oil_gas/natural_gas/special/ngresources/ngresources.html

US Energy Information Administration (EIA). (2011, April 5). *World shale gas resources: An initial assessment of 14 regions outside the United States.* Washington, DC: EIA. Retrieved from http://www.eia.gov/analysis/studies/worldshalegas/archive/2011/pdf/fullreport.pdf

US Energy Information Administration (EIA). (2011, April 29). *Ethane prices trail other natural gas liquids: Comparative spot price movements in natural gas, crude oil, and natural gas liquids.* Retrieved May 29, 2012, from http://www.eia.gov/todayinenergy/detail.cfm?id=1170

US Energy Information Administration (EIA). (2011, September 30). *Global natural gas prices vary considerably.* Retrieved May 29, 2012, from http://205.254.135.7/todayinenergy/detail.cfm?id=3310

US Energy Information Administration (EIA). (2012, January 23). *AEO2012 Early release overview: Energy productions and imports.* Report Number: DOE/EIA-0383ER(2012). Retrieved May 24, 2012, from http://www.eia.gov/forecasts/aeo/er/early_production.cfm

US Energy Information Administration (EIA). (2012, May 9). *Henry Hub Gulf Coast natural gas spot price (dollars/mil. BTUs).* Retrieved May 16, 2012, from http://www.eia.gov/dnav/ng/hist/rngwhhdW.htm

US Energy Information Administration (EIA). (2013, June 13). *Technically recoverable shale oil and shale gas resources: An assessment of 137 shale formations in 41 countries outside the United States.* Table 6. Top 10 countries with technically recoverable shale gas resources. Washington, DC: EIA. Retrieved from http://www.eia.gov/analysis/studies/worldshalegas/

US Energy Information Administration (EIA). (n.d.). *Maps: Exploration, resources, reserves, and production.* Retrieved from http://www.eia.gov/pub/oil_gas/natural_gas/analysis_publications/maps/maps.htm

US Environmental Protection Agency (EPA), Office of Research and Development. (2011, May). *Proceedings of the technical workshops for the Hydraulic Fracturing Study: Water resources management.* Presented at US EPA Technical Workshop for the Hydraulic Fracturing Study: Fate & Transport, Arlington, VA.

US Environmental Protection Agency (EPA). (2011). *Proceedings of the technical workshops for the hydraulic fracturing study: Chemical & analytical methods*. EPA 600/R-11/066. Retrieved from http://water.epa.gov/type/groundwater/uic/class2/hydraulicfracturing/upload/proceedingsofhfchemanalmethodsfinalmay2011.pdf

US Environmental Protection Agency (EPA). (2012). *Oil and natural gas sector: New Source Performance Standards and National Emission Standards for Hazardous Air Pollutants Reviews*. Retrieved from http://www.epa.gov/airquality/oilandgas/pdfs/20120417finalrule.pdf

US Geological Survey. (2003, October). Depth to saline ground water in the United States (generalized from Feth and others, 1965, Preliminary map of the conterminous United States showing depth to and quality of shallowest ground water containing more than 1,000 parts per million dissolved solids: US Geological Survey Hydrologic Investigations Atlas HA-199). In *Desalination of ground water: Earth science perspectives* (USGS Fact Sheet 075–03). Reston, VA: Office of Ground Water, US Geological Survey. Retrieved from http://pubs.usgs.gov/fs/fs075-03/

US Geological Survey. (n.d.a). *Earthquake facts and statistics*. Retrieved May 18, 2012, from http://earthquake.usgs.gov/earthquakes/eqarchives/year/eqstats.php?mtcCampaign=10798&mtcEmail=27306046

US Geological Survey. (n.d.b). *Magnitude/intensity comparison*. Retrieved May 18, 2012, from http://earthquake.usgs.gov/learn/topics/mag_vs_int.php

Vardi, N. (2014, August 12). How one billionaire's bet on LyondellBasell turned into the greatest deal in Wall St. history. *Forbes*. Retrieved June 5, 2015, from http://www.forbes.com/sites/nathanvardi/2014/07/30/the-greatest-deal-of-all-time/

Walters, C. (2006). The origin of petroleum. In C. S. Hsu & P. R. Robinson (Eds.), *Practical Advances in Petroleum Processing*. Springer, pp. 79–101. http://dx.doi.org/10.1007/978-0-387-25789-1_2

Wynne, H., Broquin, F. D., & Singh, S. (2010). US utilities: Coal-fired generation is squeezed in the vice of EPA regulation; who wins and who loses? *Bernstein Report*. Available June 5, 2015, from http://www.docstoc.com/docs/132733407/Utilities-Coal-Fired-Generation-Is-Squeezed-in-the-Vice-of-EPA

Song Lyrics

Casey, H. W., & Finch, R. (1976). (Shake, shake, shake) shake your booty [Recorded by KC and the Sunshine Band] [7" single]. Miami, FL: TK Records.

Cash, J. (1956). I walk the line [Recorded by Johnny Cash] [7" single]. Sun Records.

Christie, L., & Herbert, T. (1966). Lightnin' strikes [Recorded by Lou Christie] [7" single]. MGM Records.

Crewe, B., & Gaudio, B. (1967). Can't take my eyes off you [Recorded by Frankie Valli] [7" single]. Philips Records.

Croce, J. (1971). You don't mess around with Jim. On *You don't mess around with Jim* [Album]. ABC Records.

DuBois, T. (1982). She got the goldmine (I got the shaft) [Recorded by Jerry Reed] [7" single]. RCA Records.

Dylan, B. (1962). Blowin' in the wind [7" single]. Columbia Records.

Dylan, B. (1964). The times they are a-changin'. On *The times they are a-changin'* [Album]. Columbia Records.

Felder, D., Frey, G., & Henley, D. (1977). Hotel California [Recorded by the Eagles] [7" single]. Asylum Records.

Fogelberg, D. (1981). Leader of the band [7" single]. Full Moon Records.

Fox, C., & Gimbel, N. (1972). Killing me softly with his song. [Recorded by Roberta Flack] [7" single]. Atlantic Records.

Jagger, M., & Richards, K. (1965). (I can't get no) satisfaction [Recorded by The Rolling Stones] [7" single]. London Records.

Joel, B. (1979). You may be right [7" single]. Columbia Records.

Knopfler, M. (1985). Why worry [Recorded by Dire Straits]. On *Brothers in arms* [Album]. Vertigo Records and Warner Brothers Records.

Lennon, J., & McCartney, P. (1965). We can work it out [Recorded by The Beatles] [7" single]. London: EMI Studios.

Lennon, J., & McCartney, P. (1969). Get back [Recorded by The Beatles] [7" single]. London: Apple Records.

Lennon, J., & McCartney, P. (1970). Let it be [Recorded by The Beatles]. On *Let it be* [Album]. London: Apple Records.

McLean, D. (1971). American pie [7" 45 RPM single]. Los Angeles, CA: United Artists Records.

Nettles, J. (2006). Stay [Recorded by Sugarland]. On *Enjoy the ride* [CD]. Nashville: Universal Music Group.

Page, J., & Plant, R. (1971). Stairway to heaven [Recorded by Led Zeppelin] [EP single 45 RPM]. Atlantic Records.

Porter, C. (1934). I get a kick out of you [Recorded by Ethel Merman]. In *Anything goes* [Musical]. New York: Harms Inc.

Ragovoy, J. (1963). Time is on my side [Recorded by The Rolling Stones] [7" 45 RPM single]. Verve Records.

Sebastian, J. (1965). Did you ever have to make up your mind? [Recorded by The Lovin' Spoonful] [7" single]. Kama Sutra Records.

Sebastian, J. (1965). Do you believe in magic. [Recorded by The Lovin' Spoonful] [7" single]. Kama Sutra Records.

Simon, P. (1969). Bridge over troubled water [Recorded by Simon & Garfunkel] [7" single]. Columbia Records.

Stevens, C. (1971). Peace train. On *Teaser and the fire cat* [Album]. A&M Records.

Stills, S. (1966). For what it's worth [Recorded by Buffalo Springfield]. [7" single]. Atco Records.

The Doors. (1967). Break on through (to the other side) [7" vinyl]. Elektra Records.

Tyler, S. (1973). Dream on [Recorded by Aerosmith] [7" single]. Columbia Records.

Usher, G., Wilson, B., & Love, M. (1962). 409 [Recorded by The Beach Boys] [7" single]. Capitol Records.

Waters, R. (1979). Another brick in the wall [Recorded by Pink Floyd]. On *The wall* [Album]. Columbia Records, Harvest Records/EMI Records, and Capitol Records.

Webb, J. (1967). Up, up and away [Recorded by The 5th Dimension] [7" 45 RPM single]. Soul City Records.

About the Author

Vikram Rao, PhD, is Executive Director of the Research Triangle Energy Consortium (RTEC), a nonprofit in energy founded by Duke University, North Carolina State University, RTI International, and the University of North Carolina at Chapel Hill. Its mission is to illuminate national energy priorities and by extension those of the world, and to catalyze research to address these priorities.

The Research Triangle Energy Consortium (RTEC), representing Duke University, NC State University, the University of North Carolina at Chapel Hill, and RTI International, was formed to have a major impact on solving the nation's and the world's technical, economical, societal, and public policy problems related to the use of energy.

While in a pro bono advisory capacity to a major nongovernmental organization (NGO), Dr. Rao became very familiar with the environmental issues related to shale gas. This added to his previous knowledge of shale gas–related technology and operations. As an organization, RTEC was increasingly viewing natural gas as a transitional fuel to reach a future dominated by renewables. Also readily apparent was the fact that the nation was deeply divided regarding the ability to safely produce the needed natural gas domestically. Since rhetoric was often overtaking knowledge, Dr. Rao decided to use his background, augmented by more recent research, to illuminate both sides of the debate with a book designed to be readable by the lay public. This is the result.

Dr. Rao advises RTI International, venture capital firm Energy Ventures, and firms BioLargo, Inc., Global Energy Talent Ltd., Eastman Chemical Company, Biota Technology Inc., and Integro Earth Fuels, Inc. He is Chairman of the North Carolina Mining and Energy Commission. He retired as Senior Vice President and Chief Technology Officer of Halliburton Company in 2008 and followed his wife to Chapel Hill, North Carolina, where she is on the faculty of the University of North Carolina. Later that year he took his current position.

Dr. Rao is an engineer by training. He received his B Tech in metallurgy at the Indian Institute of Technology, Madras, India, followed by an MS and PhD in materials science and engineering at Stanford University. He has been married for 40 years to Susan J. Henning, a professor at the University of North Carolina at Chapel Hill. They have three sons. Justin, an economist, works in New York City. Colin, a mechanical engineer, works in Houston, as does his twin brother Mitchell, a computer scientist. The Raos' aging dog, Kalu, stands ready to defend their home by lavishing love on any intruder.

Acknowledgments

George King of Apache has been a willing resource and fact- and logic-checker on the upstream aspects. Raghubir Gupta and David Denton, both of RTI International, did the same on the downstream side. But if any errors have crept in, they are to be attributed solely to me. Sara Lawrence, RTI, and Daniel Raimi, Duke University, contributed data and discourse on societal issues.

Gordon Allen read the book and tailored his pen-and-ink drawings to the message, astutely handling the suggestions of the engineer author.

I am grateful to the RTI Press production crew for putting up with a rookie author—and an engineer to boot. They allowed themselves to be taken out of their respective comfort zones until convergence was achieved. Considering that they were the pros, this was a significant concession.

Many contributed, but special citation goes to Karen Lauterbach, Brad Walters, and Brian Southwell. Brian was immensely helpful early in trashing my first attempt at an Introduction. Brad somehow found time away from his day job to edit the book in timely fashion. And Karen held us all together. The reader will judge whether it all worked.

CPSIA information can be obtained
at www.ICGtesting.com
Printed in the USA
LVHW011740040820
662391LV00014B/1582

9 781934 831076